Practical Guide to

Injection
Blow Molding

PLASTICS ENGINEERING

Founding Editor

Donald E. Hudgin

Professor
Clemson University
Clemson, South Carolina

Practical Guide to
Injection
Blow Molding

Samuel L. Belcher

CRC Press
Taylor & Francis Group
Boca Raton   London   New York

CRC Press is an imprint of the
Taylor & Francis Group, an informa business

CRC Press
Taylor & Francis Group
6000 Broken Sound Parkway NW, Suite 300
Boca Raton, FL 33487-2742

ISBN-13: 978-0-3674-5313-8 (Paperback)

Library of Congress Cataloging-in-Publication Data

Belcher, Samuel L., 1933-
   Practical guide to injection blow molding / Samuel L. Belcher.
     p. cm.
   ISBN 0-8247-5791-2
   1. Injection blow molding. I. Title.

TP1151.I55B45 2006
668.4'12--dc22   2006049387

Visit the Taylor & Francis Web site at
http://www.taylorandfrancis.com

and the CRC Press Web site at
http://www.crcpress.com

Contents

Preface

High-density polyethylene is the most produced resin in the United States and the world, and its uses are growing at an ever-increasing rate. The blow molding industry is the largest user of high-density polyethylene. There are several distinct processes employed by the blow molding industry to produce blow molded products ranging from angioplasty balloons to containers holding over 1500 gallons.

Injection blow molding is one of the main processes used in the blow molding industry. The other processes are free extrusion blow molding, injection stretch blow molding, entrapment blow molding, and stretch blow molding. Each method has its advantages and disadvantages.

This book is written to reach members of the industry, from the novice to the experienced blow molder, and explain the injection blow molding industry from conception to design, costing, tooling, machinery, troubleshooting, and daily production.

Anyone in the plastics industry, including possible investors, will benefit from this book's straightforward approach, with its use of pictures, charts, figures, and supplements to show detailed tooling and container design. Basic costing procedures can be used to determine possible investment or expansion of the injection blow molding process. Costing should include resin costs, machine costs, tooling, labor, energy, floor space, overhead, sales and administration, and, of course, profit.

I do hope you will enjoy this short but detailed book and find it useful in your daily operation. There is very little published on injection blow molding, and the people that pioneered the process have mostly left the industry; thus, I have strived to bring to this book a discussion of the industry, the history of injection blow molding, and the daily ins and outs—or dos and don'ts—that we all face daily in production blow molding plants with product design, material selection, machine selection, tooling design, advantages and disadvantages, and efficiencies to be achieved. Each chapter is written to provide you with basic knowledge of the injection blow molding industry. Chapters are devoted to the history of the industry, container design, tooling design, machinery, resins, parison design, and the limitations of the process.

Whether you are in teaching, research, production, finance, tool design, or analyzing the market, this book could benefit you.

This book is dedicated to the many pioneers of this industry, including Frank Wheaton, Jr.; Ted Wheaton; Ernie Moslo; Joe Johnson, who started JOMAR; Dewey Rainville, who started Rainville Co.; Carl Andrews; John Raymonds, Sr. , who started Captive Plastics; James Dreps; Ralph Abramo; A. Piotrowsky; G.T. Schjeldahl; Angelo Guissoni; Adveple Matteo; Alfred Boced; Walter Panas; Denes Hunkar; Jack Farrell; Joe Flynn; Gottfried Mehnert; W.H. Kopetke; D. Farkus; Louis Germanio; Larry Barresi; Hataschi Aoki; and to the many people on the production floor who keep the injection blow molding machines running and the products going into the boxes every hour of the production day.

I also thank my devoted wife, Donna, who has supported me all these years and listened to my many stories of the plastic bottle industry.

Please enjoy the book and feel free to contact me at the following address:

Sabel Plastechs, Inc.
2055 Weil Road
Moscow, Ohio 45153
Phone: (513) 553-4646
www.sabelplastechs.com

Sincerely,
Samuel L. Belcher

chapter one

History

Injection blow molding is, by definition, combining injection molding with blow molding. The parison or preform is injection molded at the first station. Once the outside skin of the parison is set up so that it will not fracture when the injection mold halves separate, the rotating table rises a set height, rotates either 120° or 90°, and then drops down to lay the core rod containing the injection-molded parison into the blow mold bottom half. Then the blow mold clamp closes, capturing the injection-molded parison, and the trigger bar inside the rotating table is activated to move forward a set distance. This trigger bar movement pushes the spring-loaded core rod body or tip to open, and air or another gas enters through the core rod and lifts the injection-molded parison from the core rod and, via the air or gas pressure, forms the hot injection-molded parison to the inside of the female cavity blow mold shape. The neck or finish (threaded or snap fit) of the injection-molded parison is formed in the injection mold station and is cooled or allowed to set up in the blow mold station. Once the blown shape is allowed to cool so that the blow mold shape is retained, the air or gas is exhausted. Once exhaust takes place, the blow mold clamp opens, and the rotating table carrying the core rod and the blow-molded product lifts to its set height and again rotates either 90° or 120° to the eject station.

From the earliest days of extrusion blow molding, it was obvious to the personnel striving to produce plastic blow molded products that small containers would be costly to produce, as there was as much off-fall or regrind as there was weight in the blown product. Thus, the early pioneers searched for a new method that would allow them to produce blow-molded plastic containers that were free of off-fall.

The blow molded plastic industry really started when Dr. Jules Montier created an underarm deodorant and sought a squeezable plastic container to hold the product. The container needed to be able to be squeezed to spray the product onto the consumer's skin. Plax Corporation produced the container, and over five million units were sold the first year. Thus the plastic blow molding industry became a reality. The "Stopette" underarm deodorant squeeze bottle molded by the Plax Corporation was the first high-volume LDPE commercial blow molded container in 1946.

Interestingly enough, Emhart Corporation, the big glass machinery producer, actually owned the Plax Corporation. In the early 1950s, Owens-Illinois purchased a 50% interest in the Plax Corporation from Emhart. Later, Monsanto purchased the 50% of the Plax Corporation from Owens-Illinois in 1957 and the remaining 50% from the Emhart Corporation in 1962. The Plax name was soon dropped, and the company became part of Monsanto. Thus, Monsanto was one of the early blow molding container producers as well as a plastic resin supplier in the 1960s. Other resin companies followed this lead, such as Dow Chemical, Union Carbide, Phillips, and Owens-Illinois (then a half owner of the former USI high-density polyethylene producer).

The earliest injection blow molding process is credited to Mr. W. H. Kopetke, who had patents issued in 1943 while he was employed by the Fernplas Company. The early injection blow molding systems were actually modified injection molding machines with specialized tooling mounted between the platens of the injection molding machine. However, the earliest and most referenced process was the system developed by A. Piotrowski, depicted in panel 1 of Figure 1.1. Other developments followed, such as the Moslo (ERNIE) method (Figure 1.1, panel 2), the Farkus (D.) method (Figure 1.1, panel 3), and the Guissoni (Angelo) method (Figure 1.1, panel 4). The Piotrowski method used a center rotating plate with a vertical split parison mold and a vertical split blow mold. However, removal of the blown product, which was ejected downward during rotation, was a hindrance to the process. The Moslo method used split molds with a mold shuttle plate that moved vertically in two directions. In this method, though, the plate shutting and ejection of blow product from the blow mold were hindrances. The Farkus method employed not only a vertical shuttle plate but also a horizontal movement to allow for the blown product to be ejected. For this method, long cycle times and the presence of several plate movements were hindrances to the process. The Guissoni method was adopted and is the basis of the injection blow molding industry today. This method employs a rotating horizontal table with three main stations for injection, blowing, and ejection. Each step takes place at the same time the other step is taking place. The JOMAR , Rainville, Wheaton, Farrell, Nissei, and Bekum early machines all adopted the Guissoni method, involving the rotating horizontal table to which the core rods were attached, with the parison injection mold, the blow mold, and the ejection tooling being stationary.

Wheaton (ALCAN), Bekum, and a newcomer, Novapax station, designed the rotating table to include four stations, for safety when checking the core rods before they transfer to the injection station. The addition of a fourth station also provided for the safety station to be used to condition the core rods to producing injection blow molding containers from polyethylene terephthalate and polyethylene naphthalate. The fourth station allowed for faster cycles because the rotating horizontal table only rotated 90° instead

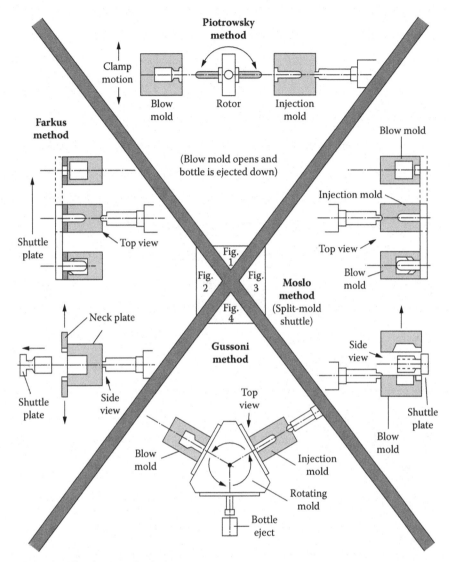

Piotrowsky method

Clamp motion

Blow mold

Rotor

Injection mold

(Blow mold opens and bottle is ejected down)

Farkus method

Shuttle plate

Top view

Neck plate

Shuttle plate

Side view

Gussoni method

Blow mold

Top view

Injection mold

Rotating mold

Bottle eject

Fig. 1
Fig. 2
Fig. 3
Fig. 4

Moslo method
(Split-mold shuttle)

Blow mold

Injection mold

Top view

Blow mold

Side view

Shuttle plate

Blow mold

Figure 1.1 The four methods.

of the regular 120°, as found on the regular three-station rotating table machines. The fourth station did add to the tooling cost, as you had to have core rods and face bars for four stations versus for three stations. The fact that the four stations can open, index, and close on a faster cycle, however, should lead all machinery producers to produce only the four-station machines. In addition, the extra station adds safety against tool damage, and the conditioning station for polyethylene terephthalate and polyethylene naphthalate is another reason to produce four-station machines.

chapter two

The process: injection blow molding

Injection blow molding combines injection molding with blow molding, thus resulting in injection blow molding. The injection phase uses either a vertical plastifier or a horizontal plastifier to take the thermoplastic resin from the plastifier's hopper and convey it into the heated barrel containing a screw for mixing and melting the thermoplastic material into a homogeneous melt, ready to be injected into a heated manifold. The heated manifold maintains the homogeneous melt and distributes it to the injection cavities of the parison injection molds. The parison injection molds contain the parison shape and are closed onto a metal core rod, which is centered in the parison injection mold. The metal core rods are mounted onto the rotating horizontal rotating table by retainers and a face bar. Once the homogeneous melted thermoplastic material is injected into the injection parison mold, the injected material is cooled so that the outside skin of the injected parison will not fracture on the opening of the injection parison mold.

The injection parison mold is split evenly into halves and one half is mounted to a stationary die plate and one is mounted to a movable die plate. The bottom half of the injection parison mold is mounted stationary in relation to the injection parison mold die plate. The heated manifold is also mounted stationary in relation to the injection parison die plate. The injection parison die stationary plate is mounted to the flat horizontal table by use of die clamps and bolts, and the blow mold die set is similarly mounted at the blow mold station. The top half or upper half of the injection parison mold is mounted to the upper injection parison die set via holding screws and key ways, and the upper half of the injection parison die plate is bolted to the movable clamp, which travels upward for opening and downward for clamping. This system is employed at the injection station and at the blow mold station. On some of the injection blow molding machines, there is a separate injection clamp and a separate blow molding clamp station. Other machines employ only one horizontal moving platen, not separate clamps for the injection station and the blow mold station. In this situation both the upper one half of the parison injection mold which is mounted to the upper die plate and the upper one half of the blow mold which is mounted to the upper die plate are both mounted to this one movable platen.

Once the injection parison mold and the blow mold have opened to their set opening, the horizontal rotating table containing the core rods that hold the injection molded parison or parisons raises and rotates either 90° (for a four-station machine) or 120° (for a three-station machine). Once rotation is complete, the rotating table drops into its set height position, and the injection parison clamp and the blow mold clamps close. This movement captures the heated parison, which is then blown off the metal core rod while a new parison is injection molded. Air or some other gas enters through the rotating horizontal table through the metal core rods, lifts the heated thermoplastic parison off the metal core rod, and forms the heated parison to the inside design of the closed blow mold cavity. The finish, or threaded area formed on the parison at the injection parison station is not blown in the blow mold but is held in the blow mold tooling to maintain its proper designed dimensions. Thus, only the body of the injection molded parison is blown or lifted off the metal core rod by the blow air pressure. The blow mold cools the now-formed product so that it retains the desired shape and dimensions. Following proper cooling of the blown product and the completion of the injection molding of the parison in the injection parison position, the clamps open vertically. The horizontal rotating table then rises to its predetermined set position, rotates either 90° or 120°, and carries the new injection molded parison to the blow mold station, the blown product to the strip or eject station, and the core rods free of product to the parison inject station. The horizontal rotating table then drops vertically downward to its predetermined position. This allows for the parison inject clamp to close, the blow mold clamp to close, and the stripper to strip the formed product from the metal core rods either for packaging or into position for secondary operations such as decorating, assembly, or treatment. The process is now complete.

This procedure continually repeats on a three-station injection blow molding machine. On a four-station injection blow molding machine, in contrast, the fourth station either allows for safety checking of the core rods to ensure that they are clean and free of any product, or there may be a vertical clamp station present, which allows for the core rod(s) to be conditioned via air chambers. This is particularly true on four-station injection blow molding machines used to produce polyethylene terephthlate or polyethylene naphthalate blown containers. The fourth station may also be used to flame heat and decorate the blown containers. Conditioning and decorating at this station will be discussed further in the chapter on processing.

This simplified explanation of the injection blow molding process as described above (pictorially explained in Figure 2.1 for a three-station machine) is just the beginning of the real injection blow molding process. The injection blow molding process begins when any plastic item to be produced begins.

The process

Stage 1–Injection

Resin is injected by the horizontal reciprocating screw into the preform cavity. The parison is formed and temperature conditioned. This is where the precise neck finish is produced.

Stage 2–Blowing

After indexing 120° counter-clockwise to the blow mold, the preform is positioned there and the mold closes. The neck finish is held while the parison is blown to the final container shape and cooled.

Stage 3–Ejection

Now the core rod, carrying the completely finished container, again indexes 120° where a stripper bar removes the finished container.

All three stages take place simultaneously with the cycle time being determined by the injection molding preform production stage.

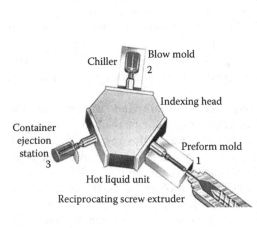

Figure 2.1

The item to be injection blow molded is first designed, whether by the customer or by a product designer, in the mold building facilities. Normally, containers holding under 8 ounces are blow molded, whereas larger containers are normally designed for free extrusion blow molding. The container design is critical with regard to the finish or top of the container and to what the critical function for this area is (e.g., to hold a closure or fitment, etc.). With the advent of childproof closures, this area became more critical to the product designs as well as to the filling and packing customer. Proper product design can save valuable grams of resin with good management of the wall thickness of the finish, the height of the finish, and the use of a flat surface area so that a top-seal membrane (such as aluminum, coated paper, or coated plastic) will adhere. Proper design of the container can save 15%–20% of the final container weight and of the cost of production. The design of the push-up in the base of the container, with regard to depth and to decorating locators, also has an effect on tool costs and cycle time. The push-up or recess in the bottom of a container serves several purposes. By recessing the bottom of a container, the designer strives to provide the container with a flat surface that will allow the filled container to set flat on a flat surface. This recess can also have an indent to allow for labeling machines to use this indent to accent the container to have the container aligned so it can be labeled as desired by the company filling the container with product. If you look at the recess or push-up you may also see the recycling code for the resin used in producing the container. These may also be the trade mark of the company injection blow molding the container and a date code may also be found in this area as well as a cavity number. The cavity number indicates what blow mold cavity was used to produce the container in question. These are limits as to how deep you may recess the push-up set by the government in the area of misleading the consumer. If you designed a deep push-up that could produce an ounce of volume from the containers volume. This is false packaging. The consumer would feel they are receiving a larger container of product versus a competitor's product due to the larger size of container.

Choosing the proper resin for a project is usually done by the company that is going to fill the container. The choice of plastic resin used is the major variable encountered in injection blow molding. In this process, the annual volume of the product and whether the user orders it on a regular basis or has a cyclical demand usually determines how many cavities will be planned for production. Naturally, the injection blow molder has to figure out what machine or machines he or she has available to produce the product.

There are several items that are of utmost consideration in choosing the injection blow molding machine to use for production. The annual volume required, the resin to be used, and the willingness of the user to pay the cost for the injection blow mold tooling all have to be considered up front so that the needed product volume can be met, proper tooling designed, and appropriate injection blow molding machine be made available.

The injection blow molding machine chosen has to be able to have an adequate injection unit for the number of grams per hour required. The machine also has to have enough injection mold clamp tonnage to allow for the chosen number of cavities and for the injection pressure that is required to fill the injection parison tooling. This measurement should include a 20% safety factor to ensure that the parison injection does not flash. Flash is excess plastic that is somewhere on the injection blow molded container, that is not specified on the container product design. It may be in the threaded area of the container, on the top surface of the container or in the recess or push-up area, and possibly on the side wall of the container at the parting line where the upper and lower injection parison mold or the upper and lower blow mold are to come to meet to form the desired injection molded parison or the desired blow molded container. Most injection blow molding machines have greater injection clamp tonnage than blow mold clamp tonnage. The injection pressure required to fill the parison mold can be approximately 3000 psi for the olefins and as high as 6000–8000 psi for engineering materials such as polyethylene terephthalate, polybutylene (PBT), polyethylene naphthalate, polycarbonate, and resin blends. The blow mold gas (normally air) is used from 80 psi up to 220 psi; thus, high blow clamp tonnage is not required. For instance, a 55-ton injection blow molding machine may have 38-ton injection clamp tonnage and only 17-ton blow clamp tonnage. Each injection blow molding machine manufacturer has its own specifications as to the injection unit, the injection clamp tonnage, the blow clamp tonnage, the fixed platen bolt pattern, and the movable platen bolt pattern for mounting the injection blow molding tooling.

Each injection blow molding machine builder also determines how large the trigger bar length is within the rotating head. The usual length of trigger bar in the rotating table determines the maximum number of cavities or core rods that can be used on this specific injection blow molding machine.

An area of concern that is critical for any injection blow molder of containers is what the open lift, index, and close time are within the machine. These steps in the process are dead time and unnecessarily add to the process time. If you have a three-station injection blow molding machine and it has a dead time of 2.8 or 3.12 seconds—open, lift, index, and close—and your competitor's machine can open, lift, index and close in 1.8 seconds, it is easy to understand which company can outproduce the other, even if they both have the same number of cavities, and who will be most profitable.

In purchasing a new injection blow molding machine, you alone—not the injection blow molding machine producer—should consider all the above options and specify the injection blow molding machine's characteristics.

chapter three

Resin (raw materials) for injection blow molding

Most of the thermoplastic materials produced in the plastics industry can be injection blow molded. The earliest plastic to be used for this process was cellulose, which was later followed by polystyrene—predominately crystal polystyrene because of its water clarity—used to package over-the-counter headache medicines. These packages still are used today; however, other resins such as polyethylene terephthalate (PET), polyvinyl chloride (PVC), polyproprene, high-density polyethylene, low-density polyethylene, comonomers, polysulfone, polycarbonate, acetal, nylon, acrylic, barex, elastomers, thermoplastic olefins, acrylic butadiene styrene, polyamide, polyethylene naphthalate, and the new metallocenes are now all being injection blow molded.

Normally, the lower–melt index resins perform well on injection blow molding machines. Nucleated materials also run well on these machines. In some resins, such as the olefins, it may be necessary to use a release agent or lubricant to prevent the heated resin from sticking to the metal core rods during the blow molding cycle, which is called having a lift-off problem. However, the release agent may interact with postdecorating and may also react with the product to be packaged. Past decorating such as paper labeling, silk screening, and hot stamping may alter the label ink used in labeling blow molded containers or adhesion to the blown container. The standard test used to check adherence of the label to the blown container is to place scotch tape on the ink or label and then peel off the scotch tape. The scotch tape should be free of any label. Plastic process can cause orientation then the plate of spaghetti would cause the strands to take on a definite pattern. In biaxial orientation the strands would be similar to a screen in a screen door with the strands aligned vertical and horizontal.

Since in the injection blow molding process the parison is only expanded in the hoop, or radially, some of the strands would align in the hoop or diameter direction only while other strands would remain non-aligned.

Injection blow molding only causes the precursor or parison to expand in one direction—the diameter or hoop. However, each thermoplastic resin that will orient during blow molding, has a temperature range that the resin must be within in order for orientation to take place. Due to the process limitations and the parisons temperature, the parisons temperature is outside the resins orientation window thus little or no orientation takes place during injection blow molding. Thus, it is extremely important for the injection blow molding bottle producer to have their customers' approval in writing and to test actual injection blow molded containers for product compatibility. Lubricants used in the injection blow molding industry are mineral oil and stearates, and of course, any colorant will also serve as a lubricant. It may also be necessary to coat the body of the core rods with a release coating such as nickel Teflon, nickel diamond, or graphite.

In injection blow molding there is minimum orientation of the thermoplastic material being used because temperature of the heated resin is normally above the orientation temperature of the thermoplastic resin. If there is minimum orientation, it will be in the hoop or radical direction because the parison is not stretched axially or vertically but is only blown radically. To understand orientation, we have to first determine if the plastic resin used to produce the blown container will orient. Thermoplastic materials used in injection blow molding that will orient are polyethylene terephthalate (PET), polyvinyl chloride (PVC), acryonitrice (Barex), polypropylene (PP), cyclo-olefinpolymere (COC), crystal polystyrene (PS) and polyamide (NYLON). High-density polyethylene (HDPE) as purchased from high density polyethylene producers does not orient.

Orientation is the fact that during processing of the thermoplastic you can cause the molecules to become aligned. In normal injection molding or normal extrusion of the thermoplastic resins there is very slight orientation of the molecules. If one considers a plate of spaghetti, there is no alignment, and the strands are random.

A good example of biaxial orientation is a PET soft drink container or Du Pont's Mylar (TM) film. Both are bivariably oriented. You cannot tear Mylar's (TM) film and for your own understanding try to destroy a PET soft drink container. Examples of unaxial orientation are sheet extrusion of various thermoplastics and blown film for covering of garments from the dry cleaners.

Certain resins or grades of resin will process easier in injection blow molding than will other grades of resin, which is also true in extrusion blow molding. It is up to the producer and the user to choose a resin that both satisfies the producer of the containers and is most effective in creating the package. You should always receive approval for two resins for use in producing the container, as there is always the possibility for a resin to be removed from production, for plant fires, for strikes, and for resin shortages. In addition, if you are producing containers for food packaging, the resin should have a no-objection letter from the Food and Drug Administration, as should any additives to the resin (i.e., colorants, lubricants, antioxidants).

Table 3.1 Resin Factors

Resin	Hot Melt Density, GMS/IN³	Shrinkage, IN/IN (×10³)	Blow-Up Ratio Container Diameter Divided by Finish "E" Diameter
High-density polyethylene	12.5	18/22	3.5/1
Polystyrene	16.11	1–3	3.5/1
Styrene aoryonitrice	16.5	1–3	3.5/1
Acrylic	17.3	1–2.5	2.5/1
Polypropylene	12.4	12–16	3.5/1
Acetal	19.5	1–2.5	2.5/1
Polyvinyl chloride	20.1	1–3	3.0/1
Polycarbonate	18.75	1–2.5	2.5/1
Acrylicpheclandiene styrene	16.7	2–3	2.5/1
Polyethylene terephthalate	21.35	1–3	4/1
Polyethylene naphthalate	21.2	1–0.25	2.5/1
Ethylene vinyl acetate	16.35	12–16	3.5/1
Cyclic obfin copolymer	13.5	1–4	2.5/1

Note: The Society of Plastics Institute has standard references for plastic bottle finishes. The "E" diameter is measured on a plastic bottle finish as the diameter across the threads of the finish.

If you are producing containers for the medical industry, the resin being used must have a Drug Master File number, and you must keep accurate records as to production dates, molds used, machines used, resin used, and lot number of the resin. You should be able to trace the container back from packaging for shipment to the raw resin received in your production facility.

Each thermoplastic resin has its own specific characteristics such as hot melt density, shrinkage, and blow-up ratio, all of which affect the product and the tooling design and production. Please see Table 3.1 for a list of these factors. A good injection blow molder will keep its own history file, to have factors such as the hot melt density and shrinkage data recorded for each container produced. This information should allow for corrections to the shrinkage factors and hot melt density based on your own production history.

Resin

PET is now second only to high-density polyethylene in use as a blow molding resin. PET's clarity, strength, color additives, availability, and acceptance in the packaging industry for everything from cosmetics to liquor, medicines, over-the-counter drugs, cough syrups, and so on has opened new markets for the injection blow molding industry—not only in the United States but also in the world market, including in China and India.

In choosing a PET resin for injection blow molding, you should pick a resin that is a slow crystallizer. The gate of the parison is the hottest area in a parison, and it is difficult to cool this gate area in the blow mold. You can design the bottom plug or push up to have cooling, but this small area of metal greatly restricts how much you can cool the gate of the parison so that it does not crystallize. Injection molded PET performs, whether they are run on single-stage stretch blow molding machines or on two-stage stretch blow molding machines, also have this problem. Cooling in the parison mold, the blow mold and the push-up is normally accomplished by having water lines designed into the molds. Using units as chilled or themalators will be set to use pumps cooling media thus the water lines designed in the molds. This removes the heat from the molten plastic and allows it to solidify in the molds.

Normally, PET will have an orientation that gives PET containers top load, drop impact, clarity, and barrier; however, in injection blow molding, a very slight hoop orientation is achieved. First, PET's orientation temperature window is approximately 190°–240°F (88°–116°C). When you are injection blow molding PET, there is no axial stretch. The blow up ratio does provide hoop stretch, but the parison is above the orientation temperature of PET, and thus very slight orientation—if any—is achieved.

The high I.V. (intrinsic viscosity) PETs, such as the 0.80-, 0.82-, and 0.85-I.V. resins, have a slower crystallization rate than do the PET resins ranging from 0.72 to 0.78 I.V. The drawbacks to using high-I.V. resins are that they are more costly for the resin producers to produce, and they require more torque and a higher melt temperature than the lower-I.V. PETs.

When designing tooling such as the injection manifold and the secondary nozzles, there cannot be any hang-up areas where the melted PET can stay or accumulate, as the resin will degrade as a result of overheating. In using PET resin in injection blow molding, you should treat it the same as if you are using PVC. PET and PVC are both shear sensitive, and both degrade when overheated. When using any resin, whether for injection blow molding, free extrusion blow molding, or injection molding, the screw in the injection unit should be designed for the particular plastic resin being used in the machine. A good screw for PET or PVC will have a compression ratio of 2.2–2.6. If olefins are being used, a good compression ratio for the screw is 3.4–3.8. It is always good practice to talk to the plastic resin suppliers and to follow their recommendations as to the compression ratio of the screw and what heat settings to use on the barrel of the injection unit, the injection nozzle, the manifold, and the secondary nozzles (if the secondary nozzles have independent heater bands).

The Wheaton patent for running PET on an injection blow molding machine is now expired. Thus, any injection blow molding machine builder can sell an injection blow molding machine for producing injection blow molded PET bottles, and any injection blow molding container producer can now produce PET injection blow molded containers.

With small injection blow molding PET containers, the inside neck diameter is relatively small, and thus the core rods diameter is small. Not many injection blow molding machines have the setup needed to run using internally cooled core rods; normally, injection blow molding PET containers have neck finishes of less than 24 mm. The method employed by companies using injection blow molding machines to produce small PET injection blow molded containers is to use a four-station injection blow molding machine, in which, at the fourth station, there is a setup whereby the core rods are cooled externally by air. Naturally, this system requires a compressor to provide air not only for blow molding the PET injection blow molded container but also for cooling each core rod. This method was used on the first Wheaton 180-ton injection blow molding machine to produce the first 8-ounce PET injection blow molded brown containers for Magic Shell ice cream topping, formerly produced by Foremost McKesson and now produced by Smuckers.

Accurate records should be kept on any plastic resin used in injection blow molding with regard to the shrinkage used in the tooling, what hot melt factor was used, and the blow up ratio. This information should be updated on the basis of your own production history.

Colorants

Color additives are a way of life in the plastic bottle industry. The color of small injection blow molded containers ranges from clear and translucent to white, to black, and even to fluorescent and metallic.

It is important that the color supplier, the container producer, and the container customer all work together. Once the resin is chosen and its melt index, or I.V., is known (for PET, polyethylene naphthalate), then the carrier that will be used to carry the color into the base resin can be chosen. If the base resin for the container has a melt index of 3 or 5, it is not uncommon for the color supplier to have a carrier that has a melt index of 10 or higher. It is not uncommon to have a high-density polyethylene or polyproprene resin as the base resin and to have a carrier for the color to be a low-density polyethylene or Liivear low–density polyethylene (L) low-density polyethylene resin. This is done to ensure that the carrier resin melts freely and disperses the colorant within the base resin. Many major companies supply their color supplier with the base resin they wish to be used as the colorant carrier. This is the best scenario for the companies, as it gives them more control of the dispersion shrinkage and of the overall quality of the finished product.

The specific resin that is being used as a carrier for the color additive purchased, plus any additives the color supplier may add, should always be known, as should the carrier melt index and melting point, to ensure that the carrier is compatible with the base resin's melt index and melt temperature.

If you add colorant to PET, you must ensure that the carrier can withstand a drying temperature of PET (540°F) for 5 hours and a melting temperature as high as 560°F. The colorant should not act as a nucleating agent in the PET base resin, as, for example, titanium dioxide does when used as a color additive for PET. Titanium dioxide can cause crystallization within the PET container, thus affecting drop impact and causing the container to embrittle.

In choosing a PET for injection blow molding, you should choose a slow-crystallizing resin. A good homopolymer PET to process is Eastmans 9663. Several copolymers can be used, such as M&G 8006 or Eastmans 9921W. It is always in your best interest to work together with your resin supplier, your colorant supplier, and your customer at the beginning of your project.

Regrind

One of the advantages of injection blow molding is that there is no moille or tail to regrind, as is found in extrusion blow molding. Also, good injection blow molders will have consistently high efficiencies (95%–98%), and thus minimum regrind. However, injection blow molding does have regrind as a result of normal quality problems such as black specks, streaks, short shots, weight variation, dimensions, and so on.

The manner in which this regrind is handled is of importance to any injection blow molder. However, regrind in most plastic plants is only given attention because it is produced from containers that, for some reason, are not packaged or sold. "All too often, grinders are purchased on the basis they are needed, but don't pay too much," is the comment made to the purchasing department. This decision is a costly and major dilemma in any plastic plant that grinds off-fall or scrap parts.

Every plastic plant that does any regrinding should use separating-type grinders. There are different types of separating-style grinders, manufactured by most of the companies that manufacture and sell plastic grinders. Separating-style grinders remove the fines, which are of a different molecular weight than the base resin pellets and do not melt at the exact same temperature as the base resin pellets. The fines are then gathered in a bin or bag for later disposal. Fines are in new plastic resin as received. You can work with your plastic resin supplier to set limits on fines in your new resin—this is the first step to ensuring that your parisons are consistent and void of black specks, gels, streaks, and starbursts. It is very good practice to personally walk into any plastic plant producing plastic products and visit the grinders. There will be fines on the grinders, the floor, and other equipment. You should eliminate the maximum amount of fines possible. Two important aspects of purchasing a grinder are the screen size and the knife angle. These two features determine the size of your regrind and also the amount of fines generated during the grinding of off-fall or scrap product. Fines are dust like particles that are present due to the plastic resin being cut into pellets or flakes and also due to the function of the pellets or resin rubbing against each other in transport converging and grinding. If you want to see fines for yourself you can open a bag of new plastic resin and place your hand down into the resin and then remove your hand and your hand will have dust like particles covering portions of your hand. These are fines.

The feed throat and the screw of your machine were designed to pick up and feed plastic pellets; they were not designed to pick up and feed irregular chopped or ground plastic scrap. The feed throat and the screw of any machine you may purchase for processing thermoplastic pellets are designed to feed and pick up the plastic pellets supplied by plastic resin producers. The plastic pellets as supplied by the plastic resin producers are of a standard size, i.e., 1/8 inch cube, 1/8 inch sphere, and are very uniform due to the plastic resin suppliers manufacturing process. The machinery producers that supply the injection molding industry, and the extrusion blow molding industry designed the feed throat and screw of their respective machines to handle these plastic pellets. They were not designed to pick up and feed irregular chopped or irregularly ground plastic scrap. The feed throat should provide a uniform, consistent feed to the screw. This cannot happen when the pellets are one size and your regrind a variety of sizes. This will cause surging in the screw, bubbles in parisons, irregular shot weights, and irregular packing of the parison. In choosing a grinder, the chips that result should be uniform and in close proximity to the size and shape of your base plastic resin. In large part, for blow molding and thermoforming, the plastic producing plant's profit or loss can often be directly attributed to how they handle their regrind.

It would be unfair to stop discussion of regrind at this point. In most plastic production plants, there are blenders sitting atop the feed throats of the injection machine. The blenders are blending the base plastic resin, the color additives, and the regrind to the feed throat. The accuracy of this blender must be checked and set for consistent feed of each item to be blended to the feed throat of the injection machine. I have often stated to blow mold container producers that I will gladly take just 10% of the increased efficiency in their production plant as my pay once the grinders and blenders are corrected. No one has ever taken me up on this challenge.

Several plastic bottle producers (Graham Packaging) repelletize their regrind before feeding it to their production machines. This is possibly the best method for consistent production; however, not many plastic bottle productions use enough of the same color and resin to dedicate a pelletizer to this area. Another solution is to utilize a central blending of the base plastic resin and to regrind and feed from this central blend area to the blow molding machines (Graham Packaging also does this).

chapter four

Advantages and disadvantages

Injection blow molding is normally the preferred production method for containers that contain 8 ounces of product (0.24 L) and less. Table 4.1 compares extrusion blow molding to injection blow molding (IBM) for container production.

Each system has applications for which it is the only feasible method for the production of a specific container. There are a number of factors that can influence the decision, but ultimately the unit container cost and quality will determine the choice of the process.

Typically, the higher tooling cost, increased production efficiency, and improved container uniformity of injection blow molding are compared with the lower tooling cost, lower output, and fairly rapid availability of extrusion blow mold tooling. The efficiency of IBM should be 93%–98% continually. The efficiency of your process may be calculated. If you know how many cavities on the machine you are using, the cycle time to produce the part, and how many hours the machine was in production, then you can figure how many plastic parts the process should have produced versus how many plastic parts you actually packed out in the specific time. For example, if the machine was scheduled for 24 hours and was using six (6) cavities running on a ten (10) second cycle. You could calculate:

$$\frac{3600 \text{ seconds per hour}}{10 \text{ second cycle}}$$

then you would have 360 cycles per hour times. The number of cavities which is six (6) and multiplying six times 360 would yield 2,160 parts per hour and on a 24 hour basis you should produce $24 \times 2,160$ or 51,840 plastic parts. If you only packed 49,856 acceptable plastic parts during the 24 hour running of the machine, the efficiency would be 96% (49,856 divided by $51,840 = 0.9617$ or 96%). Extrusion blow will average 85%–92%, depending on the container and trim requirements. Both the IBM machines and the extrusion blow molding machines may be fully processed controlled. Most plastic machinery in today's market now use a programmable logic controller that allows the operator to set all the machine operating parameters. It is usually a touch screen that has a standard machine operational format set by the machine manufacturer. It allows the operator to enter all the heat set points for the heater bands, the rate of injection, screw speed, screw back pressure, open and close time of the clamp, injection time, cooling time, blow time, overall cycle time, and other critical conditions for the safe machine operation.

Future

Energy costs are rising worldwide, and this trend will only continue. With the move of injection molding to all-electric machines, all-electric machines are the future of the process. (ALCAN) Wheaton has several IBM electric machines in production at present. The cost of an all-electric machine is higher than that of one of the standard injection blow molding machines produced by JOMAR or Uniloy Milacron, or even by (ALCAN) Wheaton themselves. However, the cleanliness, quietness, and faster overall cycle time of an all-electric machine are combining to move the total IBM machine market to comprising entirely all-electric machines.

The coinjection of parisons pioneered by polyethylene terephthalate will follow into the IBM market. This change will enable markets that have traditionally belonged to metal, composite, or glass containers to make a switch to IBM container producers. Many markets that presently use glass, metal or composite materials are striving to switch to polyethylene terephthalate (PET), however, the PET does not provide adequate protection from oxygen ingress or egress, thus another plastic resin as ethylene vinyl alcohol or nylon is injected with the PET to form a multilayer container examples include catsup bottles, beer bottles, etc.

We are in a world market today, with our goods traveling longer distances than ever before. Shelf-life requirements are necessary to ensure that the packaged product arrives at the final customer (the public) in a safe and secure form. Thus, container requirements are becoming increasingly stringent to meet the demands for them, coming from all parts of the world.

Table 4.1

	Extrusion Blow Molding	Injection Blow Molding
Scrap or off-fall	From 5% to 40% off-fall that must be reground because of the tail and moille. Handleware increases the amount of regrind.	Minimum off-fall. Normally less than 1% caused by rejects and start up. All containers are finished in the mold except for unusual shapes as automotive boots.
Tool cost	Extrusion blow mold tooling can be 30%–40% less than injection blow molding tooling. It is normally cost effective when making fewer than 300,000 containers per year.	Tooling is precise and expensive because of the need for an injections station and blow station molds, core rods, face bars, and die sets. The high cost of the process may be justified by its efficiency (92%–98%) and basically little off-fall to be reground.
Bottom pinch off	The bottom pinch off can create a weak point or stress crack area. Trim is necessary to remove the tail.	There is no pinch off.
Bottom push-up	The push-up can be difficult if too high. If too low, the pinch off may cause the container to not set flat.	Large design freedom. The bottom plug to produce the push-up can be water cooled and can be retractable for deep push-up designs.
Clarity	There is always the possibility of die lines or extrusion lines.	Containers may be water clear with no die lines, as in extrusion blow molding.
Neck finish	The neck area or threaded area may be compression molded via the blow pin, or the finish area may be blown. The finish will require trimming and, also, possible reaming. Regrind is necessary for grinding the moille and reaming if this is accomplished.	Excellent neck finish or threaded area because this area is injection molded. Undercuts are used, as in producing mascara containers. Dimensions can be held, as in injection molding.

(Continued)

Table 4.1

	Extrusion Blow Molding	Injection Blow Molding
Wide-mouth containers	Wide diameters of containers can be produced; however, posttrimming is necessary, and regrind or off-fall percentage can be in the 35% range.	The diameter of wide-mouth containers is limited because the machine opening is 5 or 6 inches (127–152 mm). However, the neck finish is excellent because it is injection molded.
Inside neck tolerance	If the finish is produced via compression (calibrated neck finish), under cuts are possible. The inside dimensions will be difficult to hold, and high tolerances may be required.	Because the finish is injection molded, excellent tolerances are attained, such as those used in producing underarm deodorant roller jar containers.
Special shapes for safety closures	Tolerances are difficult to hold because the safety feature is blown. The design possibilities for safety features are limited.	Offers a wide range of safety designs because the neck finish is injection molded.
Handle ware	This is at present the only method to have hollow handle ware produced. High off-fall necessitates more regrind, which can cause the regrind to be 35% or greater because of the handle design and location.	At this time, handle ware is not possible.
Machine casts	If medium-size containers are produced, the machine cost will be very close to injection blow molding.	Cost per thousand per hour improves with higher cavitation, such as 8, 10, 12, 14, and so on. Cycle times are faster on injection blow molding, offsetting the higher tooling cost.

(*Continued*)

Table 4.1

	Extrusion Blow Molding	Injection Blow Molding
Set-up time	This depends on personnel and how setups are handled in the production plant. Normal consideration is two men for 2 hours. However, it can also take 8 hours or longer.	This depends on personnel and how setups are handled in the production plant. Normal consideration is two men for 2 hours. However, it can also take 8 hours or longer.
In-mold decoration	This is possible but must be planned ahead. It is usually a costly addition to the machine and decreases the overall production efficiency. It will also slow down the cycle of the machine.	If a three-station machine is running, in-mold labeling will be difficult and expensive. However, if a four-station machine is running, in-mold decoration is quite feasible and does not affect overall cycle time. Overall production efficiency may be affected.
Coextrusion	Extrusion of containers containing different resins is possible and is done every day in industry. However, the machines are expensive.	Coinjection of the parison is possible and is already being done in select markets. However, the machines are expensive, and efficiency is reduced because of the coinjection.
Flame treatment	The need to flame treat containers for postdecorating is not a problem and can be fully automated at relatively low cost.	The blown containers can be flame treated as they transfer to the ejection station or at the ejection station, whether on a three-station injection blow molding machine or a four-station injection blow molding machine.

(Continued)

Table 4.1

	Extrusion Blow Molding	Injection Blow Molding
Oval containers	A wide range of oval or flat oval designed containers is possible if the tooling is designed properly.	Flat ovals and oval containers are possible. The injection molded parison can be ovalized and designed for good container wall distribution. Injection blow molding oval containers usually have better wall distribution than comparable injection blow molding produced oval containers.

chapter five

Container

The heart of injection blow molding lies in the parison. Each container shape has its own particular and unique core rod and parison design. If the initial design of the parison and core rod is not suited for the particular bottle to be molded, nothing can be done to the process to ensure that good bottles will be produced. Redesign and rework or replacement of the tools becomes a very expensive necessity.

The first step is to evaluate the container's shape. The container must not be too tall in relation to its neck finish diameter. The shape is checked by determining the ratio of the core rod length to the finish "E" diameter. In general, the L/D (length to diameter) of the coil rod should not exceed 12/1. If this ratio is achieved, core rod deflection will be minimized, and uniform material wall distribution will be maintained. The use of a programmed injection speed for the preform cavity fill can extend the L/D maximum guidelines (Figures 5.1 and 5.2).

Second, the container should have a blow-up ratio of 2.5:1 or less for optimum processing. The blow-up ratio is the container body maximum diameter divided by the finish "E" diameter. Larger blow-up ratios are possible; however, the greater the blow-up ratio, the greater the chance for nonuniform wall distribution. In exceptional cases, ratios up to 3.5:1 can be achieved. These containers usually exceed the recommended blow-up ratio in only a portion of the container.

Oval containers also must be checked to see whether they fit within acceptable ovality ratios—the ratios of width to depth. Satisfactory containers can be produced within an ovality ratio of 1.5:1, with the use of circular cross-section parisons. At ratios of 2:1, round core rods and round preform molds should be used. Ovality ratios greater than 2:1 usually require oval parison molds to minimize weld lines, and increased tooling and development costs must be recognized as possibilities as the ovality ratio increases. The upper limit of the ovality ratio is somewhere near 3:1.

The basic shapes of containers are round, square, oblong, or oval, as depicted in Figure 5.3. To begin to design a specific container, the designer and the customer (who may be one and the same in some instances) have

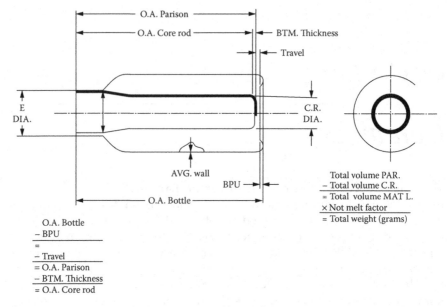

O.A. Bottle
− BPU
=
− Travel
= O.A. Parison
− BTM. Thickness
= O.A. Core rod

Total volume PAR.
− Total volume C.R.
= Total volume MAT L.
× Not melt factor
= Total weight (grams)

Figure 5.1

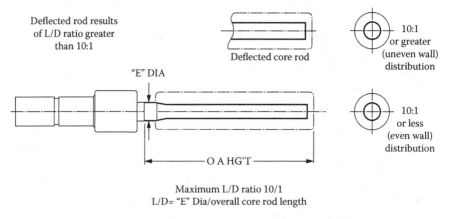

Deflected rod results
of L/D ratio greater
than 10:1

Deflected core rod

10:1
or greater
(uneven wall)
distribution

"E" DIA

O A HG'T

10:1
or less
(even wall)
distribution

Maximum L/D ratio 10/1
L/D= "E" Dia/overall core rod length

Figure 5.2

to decide on the plastic material to be used, the physical size and shape of the desired container, the neck finish size, the shape, the threads, the undercuts, the use of a locking ring on projection and possibly a rachet to indicate the use of a tamper evident closure, the parting line location, the surface finish, the push-up, the necessary engraving, the overflow capacity, the fill point and target weight of the container. Minimum wall thickness must be specified as well as the location of the minimum wall thickness allowed.

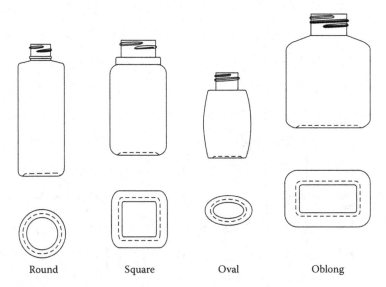

| Round | Square | Oval | Oblong |

Figure 5.3

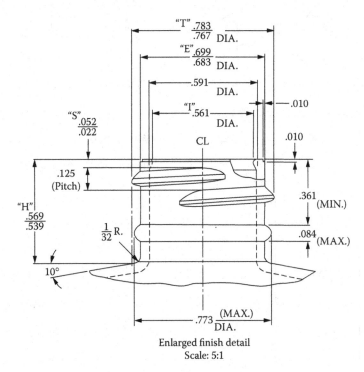

Enlarged finish detail
Scale: 5:1

Figure 5.4

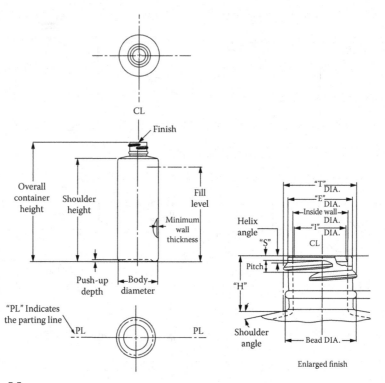

Figure 5.5

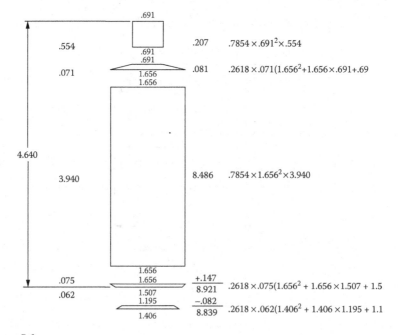

Figure 5.6

A 4-ounce (0.118-L) round container made from high-density polyethylene with a specified neck finish of a M20-SP400, as shown, will be used as an example here. Figures 5.5 and 5.6 are preliminary layouts that can be followed for any container design. The figures use the normal nomenclatures used in container design.

Basic design begins by sectioning the desired container into known geometric shapes so that you can calculate the volume in cubic inches. In Figure 5.6, all the volumes have been calculated and add up to 8.921 cubic inches. However, the push up, which has 0.082 cubic inches, must be subtracted, and thus the total displacement of our 4-ounce (0.118-L) round container, with this particular finish, will be 8.839 cubic inches (224.51 mm). The dimensions used for the container's total displacement are all outside dimensions. They will be used to calculate the proposed container's weight and volume. For example, a 4-ounce (0.118-L) container will have 7.218 cubic inches (1 ounce = 1.8047 cubic inches; thus, 1.8047 × 4 = 7.218 cubic inches).

Usually, each container is designed with its overflow capacity, which is the total liquid volume of the container to the exact top of the container's finish, in mind. The volume that falls between the overflow point and the designated fill point is referred to as the headspace. The amount of headspace desired can be a function of the type of filler, the nature of the product to be packaged, whether there are internal plug fitments, and the expansion of the product to be packaged. In calculating for the container's volume and weight, an average wall thickness must be specified.

The process begins by using an average wall thickness of 0.020 inch (about 0.5 mm). The wall thickness can then be used to calculate the fill level. If the fill level is too low, the average wall thickness can be increased, the push up can be increased, or the container height can be reduced.

Once the fill level has been established, you have also established the average wall thickness of the container. The final calculations in Figure 5.7 show that the final overflow capacity is 8.102 cubic inches, which is equal to 4.489 ounces (0.132 L; 8.102 cubic inches divided by 1.8047 cubic inches).

We can now subtract the total container's displacement (8.839 8.102 = 0.737 cubic inches). Given that the container is to be produced using high-density polyethylene, which is listed as weighing 15.62 g per cubic inch, the proposed container's weight can now be calculated (0.737 × 15.62 = 11.51 g).

The final container can now be fully drawn and dimensioned, including overall height, push-up height, overflow capacity, fill level, weight, and average wall thickness. The size in cubic inches of the container's gram weight is used in designing the parison for the container.

With the use of computer-aided design, all the SPI finish dimensions and weights can be put in a computer. Normal container design is completed on the first day of design, excluding the finish; the finish is added to the container body. Through the use of CAD (computer aided design) all the SPI (Society of Plastics Industry) finish dimensions and weights can be placed in the computer with a file number assigned to enable the designer to go to this file for the specific finish that is to be used with the container design.

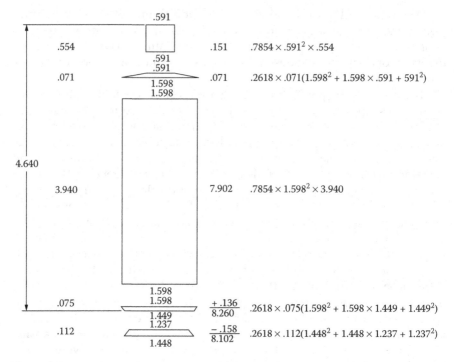

Figure 5.7

Normal container design is first completed and then the specified or desired finish is added to the container design. With the use of computers and the many programs available to the designer and to the mold builder, it is possible to have prototype molds for sampling available in as little as 10 days. With these advances in technology, there is no need to make either a model or hard-copy drawings.

In designing for a rounded square container, an oval container, or an oblong container, the basic parameters followed when designing a round container are followed. Additional calculations are necessary when using the old layout methods, versus using a computer. Figures 5.8, 5.9, and 5.10 all depict the layout of the rounded square container. In Figure 5.11, you must lay out the start of the threads along the parting line to prevent an undercut in the thread area. If you add all the volumes of Figure 5.11 together, at the total is 13.292 cubic inches. Once the push up is subtracted, the total displacement is 13.140 cubic inches.

Figure 5.10 can be used to arrive at a volume of 11.714 cubic inches. Converting this total to ounces, the volume will be 6.490 ounces (0.192 L). By subtracting the volume from the total displacement and multiplying by 14.75, as the plastic material used is polypropylene, the weight of the designed container is 21.03 g. The final container can now be fully drawn and dimensioned, including overall height, push-up height, overflow capacity, fill level, weight, and average wall thickness.

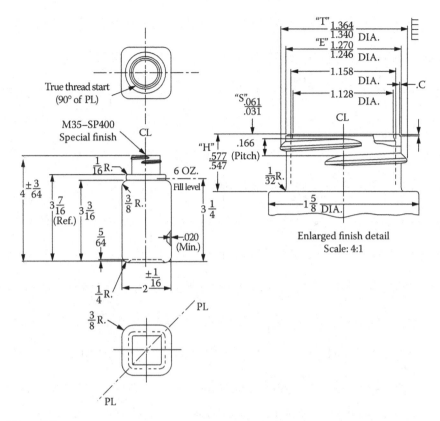

Figure 5.8

By following the above examples, the weight of the final product can be determined. Before the advent of the computer, the product to be produced had to be segmented into geometric designs for calculations. Today, the computer does all the calculations, with minimal hand calculations needed.

As in any process, there are basic rules to be followed in design. Injection blow molding is no exception. When designing a parison for oval containers, the parison is usually ovalized in the direction of the container's depth. The ratio for the maximum wall thickness in a parison to the minimum wall thickness in the parison across the parison's cross section should be less than 1.5 to prevent knit or weld lines.

A normal parison's wall thickness is 0.80 inch (62 mm) in the annular area, which is anywhere along the profile, except in the neck finish. There may be some compromises between the parison's wall thickness and the blow-up ratio. If a given container weight is to be maintained and the minimum wall thickness observed, then the parison will have to be redesigned to have less diameter, to maintain the specific desired containers weight, which will increase the blow-up ratio.

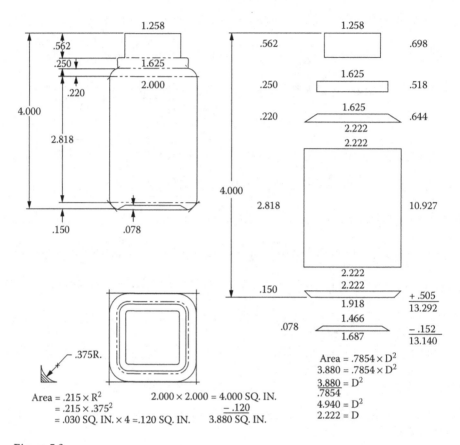

Figure 5.9

In general, a parison wall thickness of over 0.250 inch (6.35 mm) is unstable during the blow mold portion as a result of the thick parison section being inadequately conditioned in the parison injection cavity.

The parison design usually establishes the core rod length and diameter. In every case, the minor diameter of the core rod (parison inside diameter) must be smaller than the smallest diameter in the neck finish of the parison to facilitate removal of the blown container at the eject station.

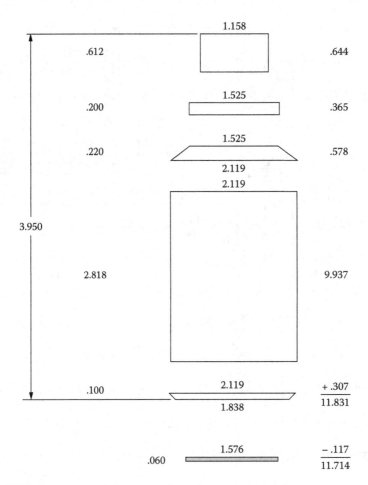

Figure 5.10

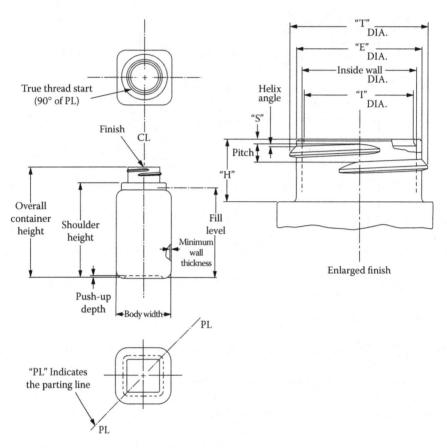

Figure 5.11

chapter six

Parison

The parison design is the "heart" of injection blow molding. The tooling is all constructed predicated on the parison. Once the specifications of the product to be produced are deemed acceptable by the designer, processor, and mold builder, the parison design can be developed. This design of the parison will establish the core rod shape and length, the material used in the core rod design, and the design's diameters.

The minimum wall thickness is usually 0.060–0.080 inches (1.5–2 mm) in the annular area—anywhere along the profile except in the neck finish. There may be some comprises between parison wall thickness and blow-up ratio in any given product to be produced. If a given weight is specified for the product, if blow-up ratios are small, or if minimum and maximum wall thickness have to be met, then the parison's wall thickness has to meet these conditions.

In parison designing, a parison layout is required. Several layouts may have to be done to enable you to visualize the blowing up of the parison into the desired product shape. The parison layout actually determines the design of the core rod because the core rod produces the inside design of the parison. First calculate the cubic inches in the parison and the cubic inches of the core rod, and then subtract the core road cubic inches from the parison's total cubic inches, leaving the total cubic inches of resin in the parison. Once this calculation is complete, multiply the cubic inches of the parison by the hot melt factor of the specific resin, resulting in the weight of the parison in grams. The hot melt factor is available from your resin supplier or from your own history file of injection blow molding processing of the same plastic resin in other products. This history file should be constantly updated.

The hot melt factor is the density of the plastic resin in the melt condition. Table 6.1 lists hot melt factors for various resins used in the injection blow molding industry. The units are grams per cubic inch.

When calculating or laying out the parison design, the projected area of the parison will be in square inches. By multiplying the number of cavities times the single parison's square inches times the planned injection pressure,

Table 6.1 Hot Density Factor to Convert Cubic Inches to Grams

Polyethylene	13.077
Polyethylene	12.5
Polystyrene	16.1
Styrene acrylonitrile	16.5
Acrylic	17.3
Polypropylene	12.4
Polypropylene alloy	12.5
Kynar	28.0
Ethylene vinyl acetate copolymer	16.35
Celcon	19.5
P.V.C.	20.1
Barex	18.35
Polycarbonate (Lexan)	18.75
PET	21.35

you will have the pounds necessary to hold the parison molds together during the injection cycle. Dividing the number by 2000 pounds per ton gives the clamp tonnage. A safety factor of 20% should be added to this answer.

An example to calculate the required clamp tonnage is as follows:

$$\frac{\text{Projected flat area of preform, sq in.} \times \text{Injection pressure, psi} \times \text{Total no. of cavities}}{2000 \; 1/\text{ton}} = \text{Minimum injection clamp force, tons}$$

For example, a parison length of 4.504 inches × 0.750 inches flat projected image of the parison times the number of cavities (8) = 27.024 square inches for a four (4) ounce round high density polyethylene (HDPE) container. The normal injection pressure for this plastic material is 3000 pounds per square inch. Thus 27.024 square inches times 3,000 pounds per square equals 81,072 pounds and dividing by 2,000 pounds per ton. This yields 40,536 tons. This is the minimum injection clamp tonnage required for this eight (8) cavity four (4) ounce high density polyethylene round container parison design.

An example as to how to use the hot factor from Table 6.1, for the specific resin to be used to determine the actual containers weight in grams is illustrated below:

For a four (4) ounce round high density polyethylene (HDPE) container, the parison's volume is 1.99 cubic inches minus the core rod's volume of 1.12 cubic inches yields 0.87 cubic inches.

Looking at Table 6.1, the hot meet factor for high density polyethylene (HDPE) is 12.5 grams per cubic inch. Thus 0.87 cubic inches times 12.5 grams per cubic inch is 10.875 grams for the weight of the desired four (4) ounce high density polyethylene (HDPE) container.

This total should be checked against the actual production container and the weight of the production container versus the calculated weight for the container should be within ≠ 0.25 grams.

With a computer, the layout of the parison and core rod and the calculation of the projected areas and volumes are quite easy, as the computer provides the important numbers to be added, subtracted, or divided on a calculator.

As previously mentioned on page 29, the Society of Plastics Industry/Society of Plastics Engineers utilizes finishes used in the industry that can be put in the computer as standards, so that when designing the body of your container, you can merely add the desired presaved finish to the container's body, simplifying the design process even further.

Hand calculations are now obsolete, as computers can now transmit the parison's layout and all tooling requirements to the mold shop without any hard drawings. The advent of solid modeling on the computer has all but eliminated the future of the model maker. Tooling that normally took 16–18 weeks to finish can now be completed and sampled in fewer than 10 working days.

Figure 6.1 below shows a parison layout for a four (4) ounce (0.118 Liter) round container. Normally, using the computer, the parison cavity and the core rod are designed in conjunction with one another.

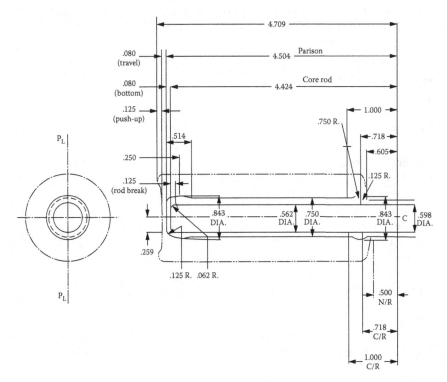

Figure 6.1

The length of the parison is the length of the finish added to the length of the core rod from the end of the neck ring length plus the thickness of the plastic measured across the bottom of the core rod. In Figure 6.1, N/R is neck ring, and CR is core rod. The travel as noted in Figure 6.1 is 0.080 inches. This is the distance the core rod shoulder or tip will move toward the bottom of the desired container to allow blow over to enter and blow the parison from the core rod to produce the desired blown container. This travel is normally set to be 0.035 inches. Shrinkage of approximately one (1) millimeter has been added to the overall parison's length. All thermoplastic materials used in blow molding have shrinkage. This happens when the hot parison is blown into a container shape in the blow mold. Once the blown container exits the blow mold and is allowed to cool to ambient temperature, the heat exits the container, and the container will shrink as it cools.

Table 6.2 below lists the plastic resin factors for the thermoplastic resins used in injection blow molding.

Table 6.2 Resin

Resin	Hot Melt Density grams per cubic inch	Shrinkage inch per inch (x10³)	Blow-Up Ratio container Diameter divided by Finish E" Diameter
High Density Polyethylene (HDPE)	12.5	18/22	3.5/1
Polystyrene (PS)	16.11	1-3	3.5/1
Styrene Aoryonitrice (SAN)	16.5	1-3	3.5/1
Acrylic	17.3	1-2.5	2.5/1
Polypropylene (PP)	12.4	12-16	3.5/1
Acetal	19.5	1-2.5	2.5/1
Polyvinyl Chloride (PVC)	20.1	1-3	3.0/1
Polycarbonate	18.75	1-2.5	2.5/1
Acrylicpheclandiene Styrene (ABS)	16.7	2-3	2.5/1
Polyethylene Terephthlate (PET)	21.35	1-3	4/1
Polyethylene Naphthlate (PEN)	21.2	1-.25	2.5/1
Ethylene Vinyl Acetate (EVA)	16.35	12-16	3.5/1
Cyclic Obfin Copolymer (COC)	13.5	1-4	2.5/1

In reality, the mold designer and the mold builder will add shrinkage to the steel dimensions when they design and build the molds. The steel dimensions include shrinkage. This allows for a slightly large plastic part to be produced, however, when it shrinks, the product will be to the proper dimensions.

In a totally new product, in which the plastic material to be used has not been run before and the article is new, such as an elastomer for front-wheel-drive plastic bellows, it is standard practice to produce a unit cavity and thus learn the behavior of the plastic material, shrinkage, possible parison design limitations, possible blow factors, and hot melt factor. Having a unit cavity trial will save valuable man hours in the tool shop and on the production floor.

In designing the parison tooling and core rod tooling, the designer should always plan for revisions. This means thinking ahead in terms of leaving enough metal to be removed if necessary to possibly add weight to the parison, which happens for various reasons. It is far easier to remove metal than to add it on any tooling.

In instances of flat oval or irregularly shaped containers, it may be necessary to shape the outside of the parison. The shaping of a parison is always done in the parison mold. In reviewing the parison mold design, there should always be a discussion on parison shaping. Figures 6.2 and 6.3 show a shaped parison for producing a rectangular bottle and a Constant Velocity Joint (CVJ) boot from an elastomer for the auto industry.

This type of shaping should only be completed after a unit cavity has been produced and sample parts have been produced. By placing concentric

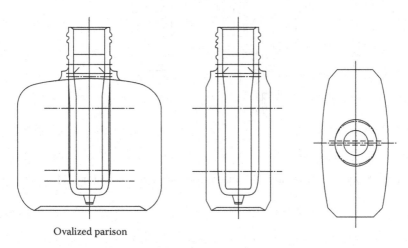

Ovalized parison

Figure 6.2 Rectangular bottle with ovalized preform.

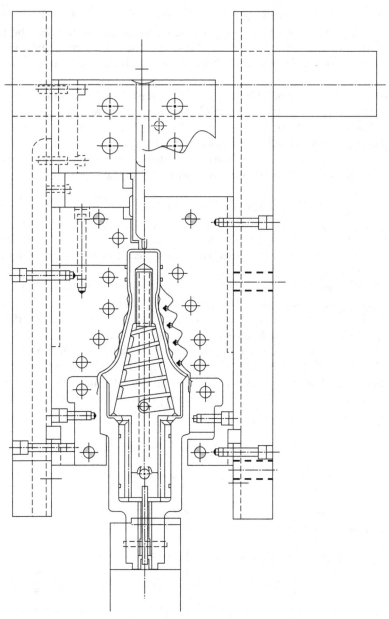

Figure 6.3 Shaped parison and thermally controlled rod core for CVJ automotive boot.

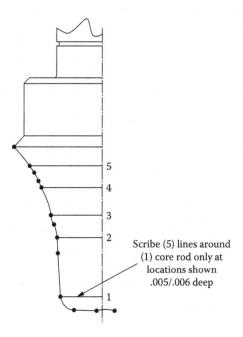

Figure 6.4

rings on the parison cavity or the core rods and studying the blow pattern of each product produced, where the material is moving and the resulting thickness can both be determined. These concentric rings can be scribed on the core rod in the mold shop per the parison core rod drawing furnished to the mold shop (see Figure 6.4).

chapter seven

Core rod design

The parison design usually establishes the core rod outside dimensions and shape, plus its length and diameter. The major diameter of the core rod (parison inside diameter) must be smaller than the inside neck diameter of the parison to allow for removal of the blown product at the eject station. Once the approved product drawing is provided to you, the processor, and the mold design/builder, the core rod basic design has been set. The overall length of the core rod is the depth of the parison design, less the bottom thickness of the parison. The head combined with the body gives the total length of the core rod to be produced.

In designing the core rod, the first question should be, How does the plastic material blow or lift off the core rod? Is the plastic material a top blow–, middle blow–, or bottom blow–type resin? You must always consider that any plastic resin used in injection blow molding is like bubble gum. The only plastic resin that does not follow this axiom is PET (polyethylene terephthalate). PET is the only thermoplastic resin that self-levels as it blows. In essence, it behaves similar to a rubber balloon, and the section that is thin or hot will normally blow first; however, as soon as this section starts blowing, PET becomes stronger than the material next to it, as a result of PETs orientation, so the material next to it blows, and so on. This process is called self-leveling. No other thermoplastic resin has this property. PET will normally begin to blow in the center of the parison, followed by the shoulder, and then the bottom. Thus, bottom-opening core rods are usually designed for PET. PET also needs high injection pressures (6000–8000 psi), and core rod deflection can be a major problem in injection blow molding production.

There are several choices available when designing the core rod. The core rod can be designed either to open at the top (top opener or shoulder opening) or to have a bottom or tip opening. The shoulder-opening core rod is the weakest design, and the bottom opener is the strongest. In looking at costs for producing either a top or bottom opener, the bottom opener will normally be the more expensive to produce. Costs are always a consideration; however, production efficiency should be the basis for the ultimate decision of whether to use a top- or bottom-opener core rod.

Temperature control can be designed in the core rod, whether it opens on the top or the bottom. The diameter of the core rod is really the main problem in adding thermal control core rods. There are several ideas for controlling core rod temperatures, ranging from cooling channels in the core rod body, to use of heat pipes, to externally conditioning the core rod.

L/D maximum guidelines for designing the core rod are seen in Figures 7.1 and 7.2. Figure 7.2 is a schematic that shows the parts making up the core rod and a top opener versus a bottom opener.

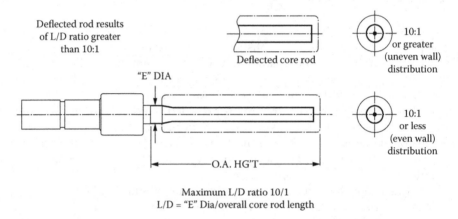

Maximum L/D ratio 10/1
L/D = "E" Dia/overall core rod length

Figure 7.1

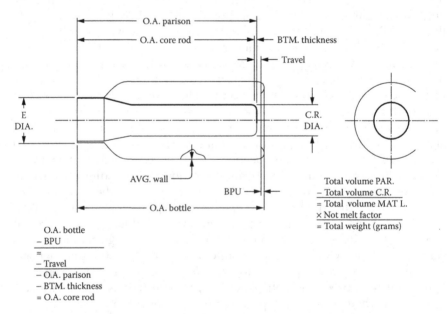

Figure 7.2
Note: O.A. – overall, BTM – bottom, CR – Core Rod, BPU – Bottom Push Up.

To determine the wall thickness of the parison in the main body, it is necessary to know what wall thickness is desired in the final blown article plus the maximum inside dimension of the desired blown article. The ratio of the inside diameter of the parison (D_1) to the maximum outside diameter of the blown article (D_2) is defined as the BUR (blow-up ratio; see Figures 7.3 and 7.4). If the desired wall thickness in the blown article is 0.020 inch and the BUR is 3 inches, then the parison's main wall body has to be 0.060 inch, as it will be expanded or reduced three times.

In an oval or oblong container or article, you must calculate the minimum BUR and the maximum BUR. In injection blowing, it is normal to ovalize the parison to produce oval or oblong products. This is similar to parison programming in extrusion blow molding.

The tip of the parison should be designed to aid in the polymer's flowing into the parison mold to reduce stress in the parison and to prevent core rod deflection caused by the injection pressure impinging on the tip of the core rod, similar to a cantilever beam being deflected by a force applied at the unsupported end.

Normally, using the largest diameter core rod that the product design permits is preferred, with the main body of the core rod being straight for two-thirds of its length, with radius of at least 0.020 inch (0.5 mm). The outside radius of the parison tip should be designed to add approximately 20% extra material, as this is the area of the parison that thins out the most when the parison is being blown. It is also the area of the parison that produces the heel of the blown container, which is normally the farthest blow point in the overall parison container design. The parison length is

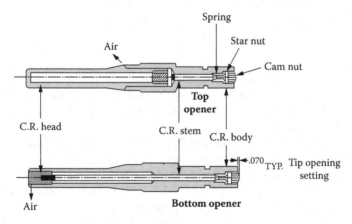

Figure 7.3

$$\frac{D_2}{D_1}$$

Figure 7.4

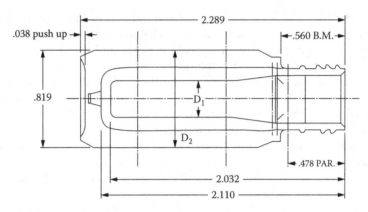

Figure 7.5

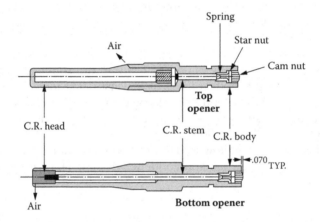

Figure 7.6

designed to clear the inside dimension of the blow mold cavity by approximately 0.060 inch (1.5 mm), as this is the distance the core rod may open to pin the tip of the parison to the blow mold and allow the air to enter and blow mold the product.

The core rod, as seen in Figure 7.5, consists of the core rod body, the core rod head, the core rod stem, the cam nut, the star nut, the spring, and—on bottom-opening core rods—the core rod tip. Various components of the core rod can be seen in Figures 7.5–7.11. Figure 7.6 shows the head for this shoulder opening production core rod.

On large products, it is very important to ensure that there are adequate air passages through the core rod head to the opening for air to exit to blow the product. Large boots, bellows, and jars can be starved for blow air if this spacing is not accomplished.

Core rod heads are usually produced from the American Institute of Steel Industry (AISI) L6 which is a very good tool grade steel with a Rockwell Hardness (RC) of 52 to 56. The core rod body can be stainless steel No. 430,

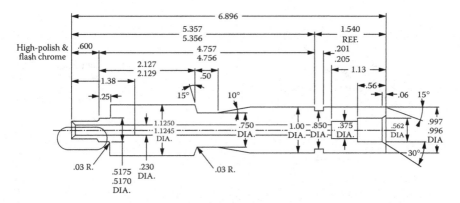

Figure 7.7

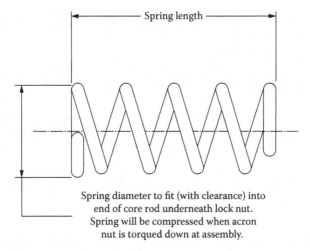

Spring diameter to fit (with clearance) into
end of core rod underneath lock nut.
Spring will be compressed when acron
nut is torqued down at assembly.

Figure 7.8 Core rod spring.

a good quality stainless tool steel. It can also be made of the L6 tool steel or Beryellium copper (BeCu) depending on the resin and product to be manufactured. The core rod body should be draw polished. Draw polishing is polishing in the direction of the plastic melt flow. In injection molding the parison, the molten plastic flows up the core rod from the core rod top. Thus the draw polish is in this direction and not accomplished in the normal circular motion as turning on a lathe. The core rod once polished should then be flash chromed. If running polyvinyl chloride (PVC), it is best to use stainless steel for the core rod and call out on the core rod drawing for an Society of Plastics Industry number one (1) polish which is a mirror polish (SPI #1). This is also true for the parison cavity. The inside or the hot inner portion of the injection manifold should also be polished to remove all tool marks and to remove any sharp edges or any area, where the melted plastic can adhere or lay which will be an area where the melted plastic will degrade since it is stagnant and will become overheated.

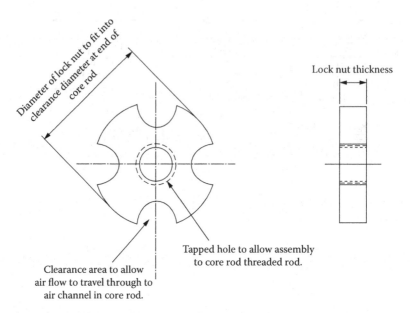

Diameter of lock nut to fit into clearance diameter at end of core rod

Lock nut thickness

Tapped hole to allow assembly to core rod threaded rod.

Clearance area to allow air flow to travel through to air channel in core rod.

Figure 7.9 Core rod washer.

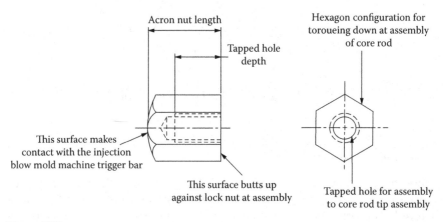

Acron nut length

Tapped hole depth

Hexagon configuration for toroueing down at assembly of core rod

This surface makes contact with the injection blow mold machine trigger bar

This surface butts up against lock nut at assembly

Tapped hole for assembly to core rod tip assembly

Figure 7.10

The mold designer should always consider the product that is being produced. In producing an article with a small inner diameter in the finish, the core rod will be small in diameter, and core rod deflection will be a factor. Normally, with finishes under 18 mm, it is advisable to use bottom-opening core rods to prevent core rod deflection.

If the plastic being used to produce the product has an affinity for metal or does not want to lift off the core rod uniformly, then it may be wise to coat the core rod body with a release coating such as nickel Teflon, nickel graphite, or nickel diamond.

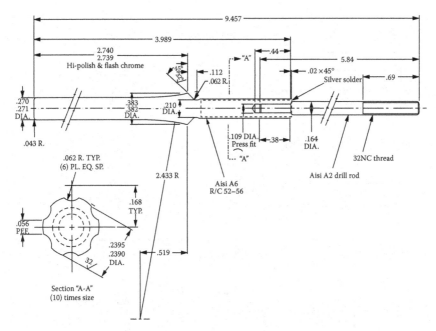

Figure 7.11

Temperature control of the core rod may also be considered if the core rod diameter has an adequate diameter. You must also consider whether the injection blow molding machine that is to be used has temperature control capability. Adding core rod temperature control capability to an injection blow molding machine in the field is not easily accomplished. Both Uniloy Milacron and JOMAR offer machines with the capability for temperature control of the core rod.

In designing for temperature control in a core rod, oil is usually used. It is advisable to design for a flow of the coolant to enter in the center and a spiral, similar to a screw thread, as the coolant returns to leave the core rod. This inner stem is best designed either to be welded into their core rod body or to be silver soldered to ensure that the core rods will not leak. Figures 7.12 and 7.13 depict temperature control core rod designs. The mold shop must test each of the core rods, once they have been produced, under pressure and temperature to ensure the rods do not leak.

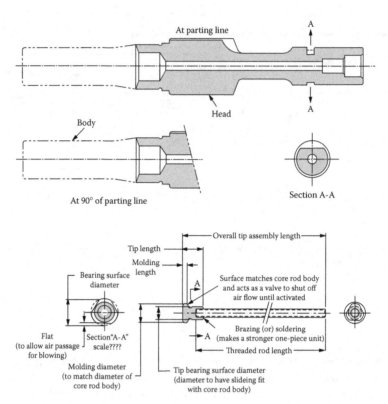

Figure 7.12 Constant Velocity Joint (CVJ) thermally controlled core rod.

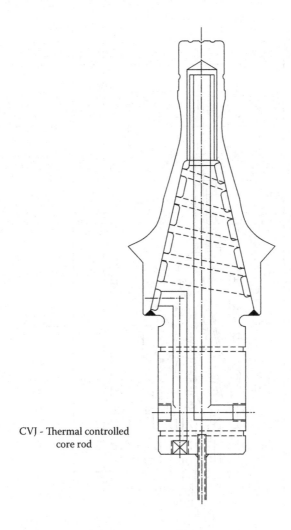

CVJ - Thermal controlled
core rod

Figure 7.13

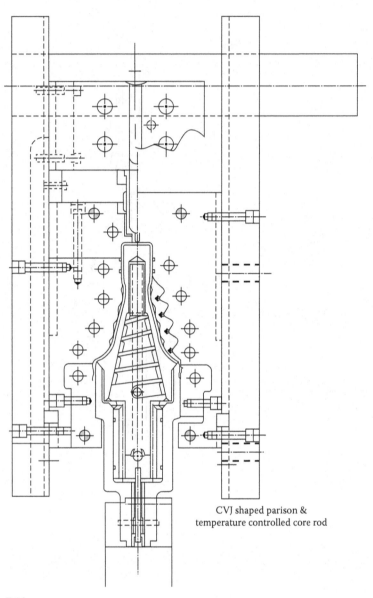

CVJ shaped parison &
temperature controlled core rod

Figure 7.14

chapter eight

Injection blow mold tooling design

Die sets

Injection blow mold tools must be machined to precise tolerances because of their critical function in the overall molding cycle. The parison molds are more critical than the blow molds because they are subjected to direct molding pressures. The core rods, similarly, are subjected to the same conditions as the parison molds and must be concentric to initiate good blowability. Because of the precise machining needed, the selection of proper tool steels to use, and the interrelations of the number of component parts making up the complete injection blow mold tooling, excessive dimensional variations between components cannot be tolerated. Figure 8.1 depicts the tooling necessary at the parison or the injection station on any injection blow molding machine. Additions to the parison injection tooling shown in Figure 8.1 include insulation under the manifold and a separate end cap to the parison cavity mold.

We begin with the parison injection mold die set. Normally, the mold builder will not produce the die set, instead purchasing the die set, including the guide posts and bushings, from companies such as DME, Mold Bases Inc., Progressive, and so on. In any injection blow mold die set design, you should use as thick a die set plate as feasible, ranging from 1.5 to 2 inches thick. On most injection blow molding machines, there is not adequate top platen support for the movable die set; thus, using a thicker die set will help defeat platen deflection, especially when there are a large number of cavities in the design.

It is a wise move to have all the die sets nickel plated to prevent rust, and it is also a wise move to have an air groove machined into the base of the stationary die set with an air inlet, so the production personnel can connect an air line to the complete injection parison tooling mold and move the mold into the machine stationary platen with ease and no damage.

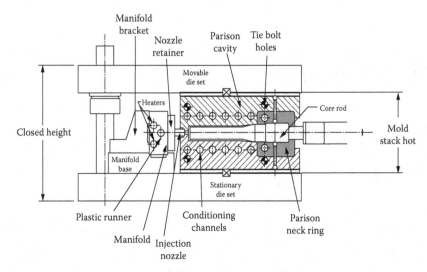

Figure 8.1 Tooling necessary at the injection station on any injection blow molding machine.

The die set should be stamped or tagged as to the tooling it holds and then dated. The size of the die set is determined by the stationary machine platen bolt pattern and the number of cavities it is to contain. Each injection blow molding machine has its own individual bolt pattern for the stationary and movable platens. The layout of these platens and bolt patterns can be secured from each individual injection blow molding machine producer.

Figure 8.2 presents a typical layout for the stationary die set. The key layout is usually cut on the center line of the die set if a cavity has its center line on the center line of the keyway. If a cavity is not going to fall on the center line of the die set, then it is better to offset the die set keyway to fall in the center of a parison cavity. The first cavity is set on the stationary die set whether it is on the center line (CL) of the die set, or offset to the right or left of the centerline of the stationary die set. This cavity is used for alignment of all the remaining cavities.

Figures 8.3 and 8.4 show the movable die set and fixed die set at the parison station. Note on the fixed die set the cut out, 1.625 inch by 1.0005/1.000, on the center line. This aligns the total mold on the fixed platen of the injection blow molding machine. Notice that all sharp corners have been removed. The fixed die set has the guideposts press fitted and is slotted on either side for allowing mold clamps to hold down the die set to the fixed platen. The die sets should both have eye bolts and straps to hold the mold together for transport and mounting in the machine.

Figure 8.5 provides the nomenclature used in injection blow molding. Closed height – normally 10 inches. This is the height of the stationary die set plus the movable die set plus the two (2) parison injection mold halves or at the blow mold station, the two (2) die sets plus the two (2) blow mold halves.

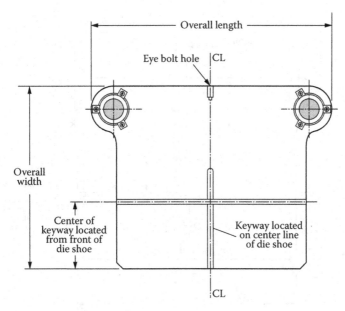

Figure 8.2 Stationary die set.

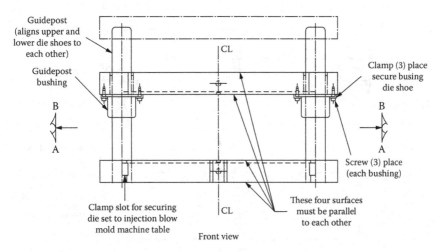

Figure 8.3 CL is Center Line movable and stationary die set at the parison station.

Figure 8.6 shows a correct setup for the parison die set, providing the normal open daylight, closed daylight, and mold shut height.

Figures 8.7 and 8.8 show the fixed die set, with holes for bolting the parison cavities onto the fixed die set. These holes can only be set once the parison cavities' number and size have been designed. The rear holes and keys are for the mounting of the manifold to this die set and can only be determined after the manifold has been designed.

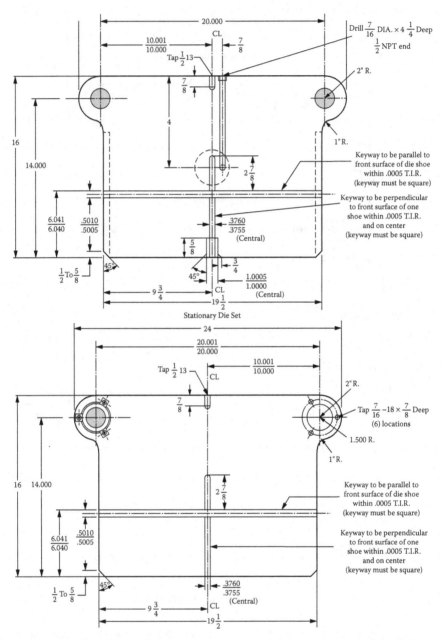

Figure 8.4 Movable die set. Dimensions for stationary and movable die sets.

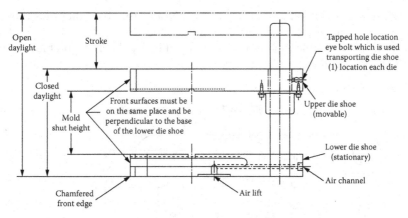

Figure 8.5 Nomenclature for die sets.

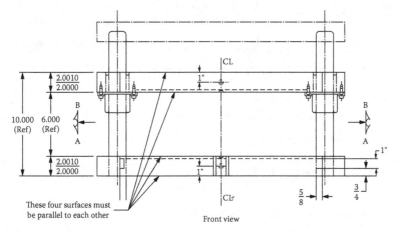

Figure 8.6 Correct set-up for parison die set.

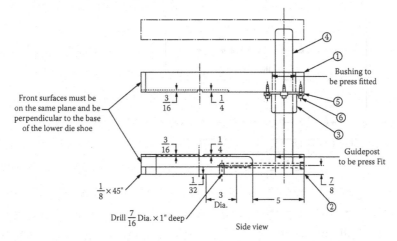

Figure 8.7 Die sets with air channel and relief on side for mold clamps.

Figure 8.9 shows the movable die set. Notice the two bolt holes that are offset. These holes are predetermined by the location of the mounting of the die set to the movable platen, which is supplied by the injection blow molding machine builder.

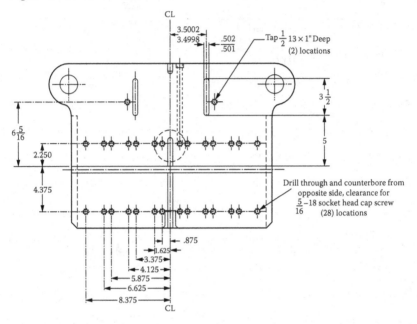

Figure 8.8 Fixed die set with dimensions.

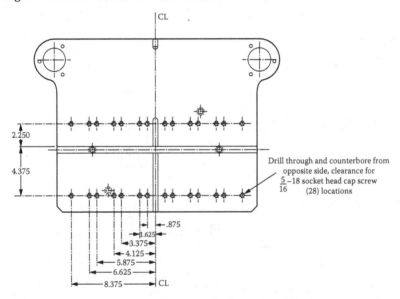

Figure 8.9 Indicates cavity mounting holes. Movable die set mold mounting dimensions.

The parison die set is critical in setup and design. Each parison die set is unique as to thickness, number of cavities, manifold design and location, and mounting in a specific injection blow molding machine.

Figures 8.8 and 8.9 depicts the bolt holes in the movable die set for a seven (7) cavity set-up. There are two (2) bolt holes that are used to bolt this die set to the clamp that moves up and down. The location of these two (2) bolt holes and the necessary thread size are provided by the specific injection blow molding machine manufacturer.

The blow mold die set is a copy of the parison mold die set. It will not be as heavy to handle as it does not contain the manifold and all the tooling as the parison molds, manifold, etc. will be tool steels. The blow molds will normally be made from quality aluminium, with steel neck inserts. They are both critical as to cavity locations, bolt patterns, cross keys and mounting into the specific injection blow molding machine.

Keys

In conjunction with the parison mold setup, Figures 8.10 and 8.11 show the key design for the parison injection die sets.

Figures 8.10 and 8.11 are dimensions and specifications for the tool shop to produce the keys to mount onto the die sets. Most mold shops will

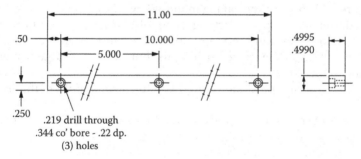

Figure 8.10 Horizontal key for use on die sets.

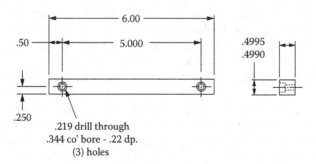

Figure 8.11 Vertical key for use on die sets.

produce the keys to be used. The keys are used to align the blow molds and the parison molds on the die sets.

Manifold

The parison die set design holds not only the injection parison cavities but also the distribution manifold to allow the melted homogeneous plastic material to leave the main injection nozzle and travel to fill each injection molded parison.

The manifold has to be designed to take the homogeneous heated plastic resin from the injection blow molding machine's plastifier and maintain its heat and then distribute the homogeneous heated plastic to each of the secondary nozzles for feeding to the injection molded preform cavity.

The manifold is mounted on a plate, and secured to the fixed injection parison die set. The manifold is supported with two (2) to four (4) support brackets to withstand the injection pressure, which can be from 3,000 pounds per square inch for the olefins to 6,000 to 8,000 pounds per square inch for polyethylene terephthlate (PET).

The manifold will have a sprue bushing that accepts the nozzle from the plastifier. It will have adequate holes for mounting the secondary nozzle retainers that hold the secondary nozzles that feed the heated homogeneous plastic to each of the injection parison molds.

Figure 8.12 shows a typical log manifold, with support brackets, secondary nozzle retainers, sprue bushing, manifold base, and the plugs at each end of the hot runner in the manifold. There will be two thermo couples located evenly from the sprue bushing to control the cartridge heaters that are in the manifold.

Figure 8.13 is a typical mounting plate for the manifold to the fixed die set.

Figure 8.14 shows a manifold support bracket. It is really up to the mold designers as to how they design the support brackets.

The sprue bushing details are shown in Figure 8.17.

When mounting the manifold to the die set, sometimes insulation is placed under the manifold. This prevents heat from the heated manifold to transfer into the stationary mounted die set and then from the die set into the plates of the injection molding machine.

Figures 8.18, 8.19, and 8.20 provide details for the insulation if the mold designer desires to use insulation under the manifold.

Figure 8.21 is another log manifold layout with dimensions for a seven (7) cavity layout.

Figure 8.22 illustrates a partial bill of materials supplied to the customer by the mold shop so the customer can buy spare parts for use with the mold design. It is a guide only as the actual bill of materials will come all the components of the mold design. Any injection blow molding bottle producer should receive a full detailed set of blueprints of the total mold design plus

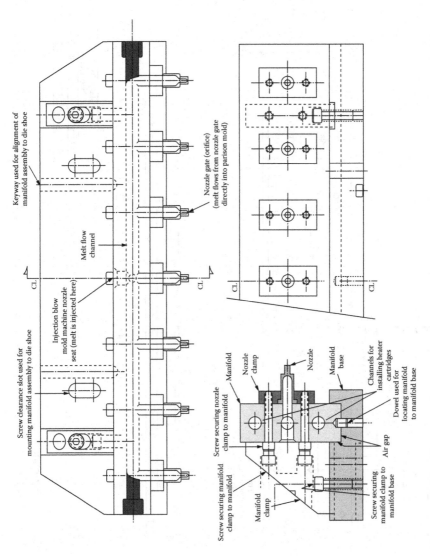

Figure 8.12 Typical log manifold with nomenclature.

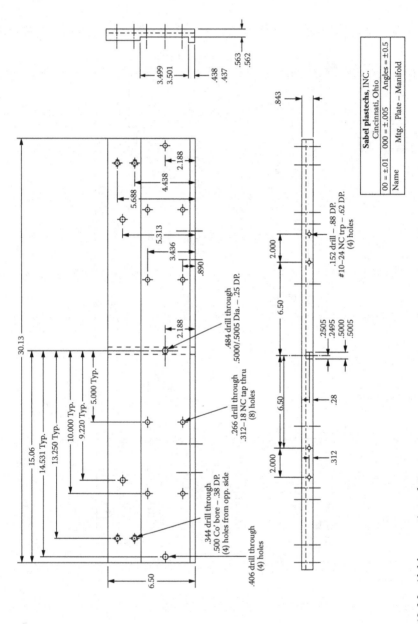

Figure 8.13 Manifold mounting plate.

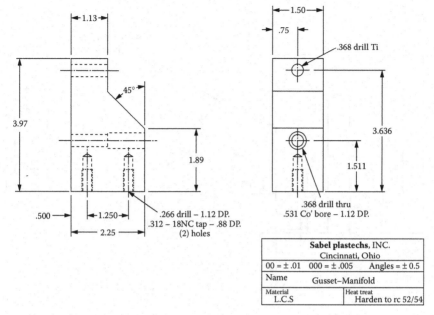

Figure 8.14 Manifold supports—two required.

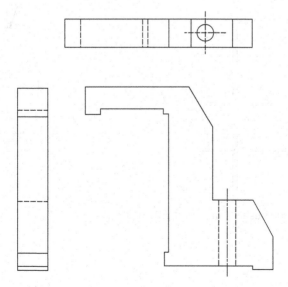

Figure 8.15 Blow mold neck insert.

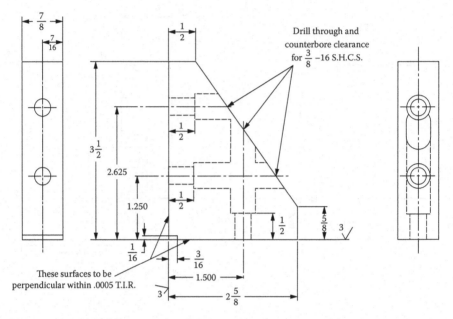

Figure 8.16 Manifold support bracket.

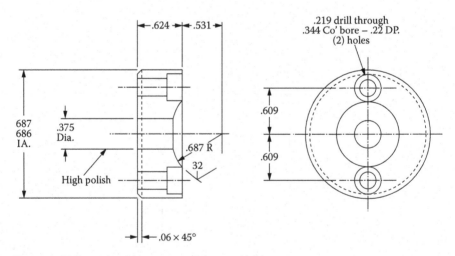

Figure 8.17 Sprue bushing mounted in manifold.

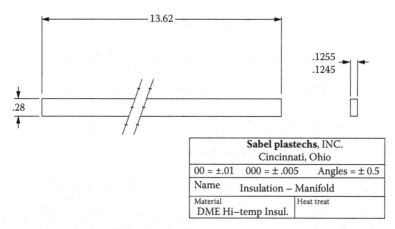

Figure 8.18 Manifold insulation.

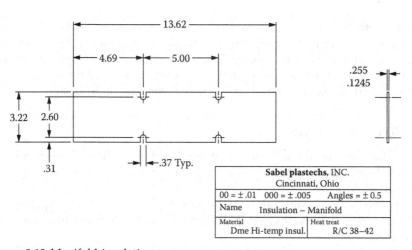

Figure 8.19 Manifold insulation.

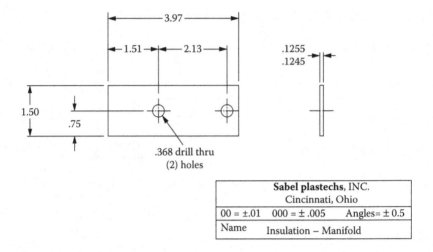

Figure 8.20 Insulation manifold.

a complete bill of materials list. This should be filed under the molds assigned number and a mold history should be maintained to include the actual purchase order the tooling's actual cost, the date of purchase, cost and date of any repairs and any particulars to setting up the tooling. Shrink factors used for each material can be taken from the mold design.

Secondary nozzles

Secondary nozzles are used to allow the travel of the heated thermoplastic material from the heated manifold and enter into each individual injection parison mold.

If you are running high density polyethylene (HDPE), polypropylene (PP) low density polyethylene (LDPE), a standard secondary nozzle is usually used.

There are many designs in use for secondary nozzles. Figures 8.23 and 8.24 are used with the proper nomenclature and usual dimensions.

However, when running other thermoplastic resins as polyvinyl chloride (PVC), polycarbonate (PC), polyethylene terephthlate (PET), cyclic olefin copolymers (COC), polysulfone and transparent nylon, a modified secondary nozzle is used as shown in Figure 8.25.

I would recommend that your secondary nozzle be of the design shown below in Figure 8.26.

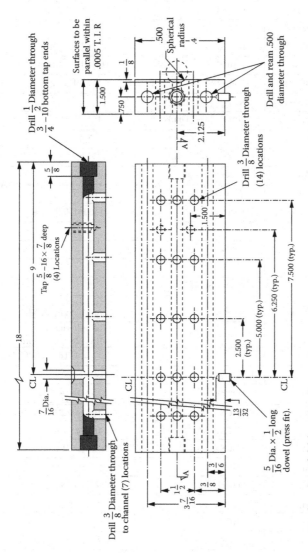

Figure 8.21 Seven cavity log manifold layout.

	Part N–	Name	Mat'l, or description
	IBM-5038-2	Assy-Manifold	~
1	IBM-5038-4	Manifold-lower	L. C. S.
1	IBM-5038-5	Manifold-Upper	L. C. S.
1	IBM-5038-6	Mounting plate manifold	L. C. S.
2	IBM-5038-7	Gusset	L. C. S.
2	IBM-5038-8	Shim	L. C. S.
2	IBM-5038-9	Insulation	DME HI-temp
4	IBM-5038-10	Insulation	DME HI-temp
2	IBM-5038-11	Insulation	DME HI-temp
2	IBM-5038-12	Nozzle (.062 Dia.)	A1S1 S 7
2	IBM-5038-12	Nozzle (.067 Dia.)	A1S1 S 7
2	IBM-5038-12	Nozzle (.072 Dia.)	A1S1 S 7
2	IBM-5038-12	Nozzle (.078 Dia.)	A1S1 S 7
8	IBM-5038-13	Guide–Nozzle	A1S1 S 7
8	IBM-5038-14	Retainer–Nozzle	A1S1 S 7
1	IBM-5038-15	Sprue bushing	4140 HT. TR. STK.
8	COMM.	Heater bond	
8	COMM.	Heater	Fast heat # CS 010323 .500 Dia. 814,00 240 V 400 W
4	COMM.		.188 spherical Dia. with .125–27 Adapter
8	COMM.		Love # 1518–22 (.062 Dia.)
14	COMM.	SHCS	312–18 × 2.25
2	COMM.	Dowel	.3750 Dia. × 1.25
2	COMM.	SHCS	# 10–24 ×.75
4	COMM.	SHCS	# 10–24 ×1.00
4	COMM.	SHCS	.312–18 ×1.75
4	COMM.	SHCS	.312–18 ×1.00
1	COMM.	Dowel	.5000 Dia. ×.62
8	COMM.	SHCS	.372–18 Dia. ×4.25
16	COMM.	SHCS	.250–20 Dia. ×3.75
4	COMM.	SHCS	.375–16 Dia. ×1.75

Figure 8.22 Typical bill of materials supplied with the tooling.

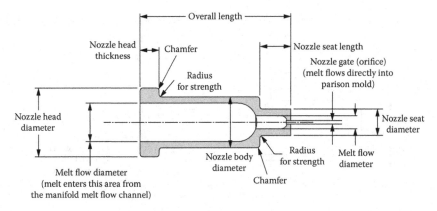

Figure 8.23 Standard secondary nomenclature nozzle.

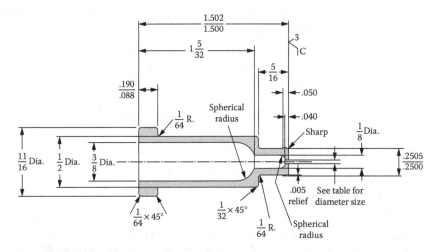

Figure 8.24 Standard secondary nozzle with dimensions.

This secondary nozzle design has no sharp corners where heated resin can be caught or hang-up and become degraded. This secondary nozzle can be used with any of the thermoplastic injection blow molding resins.

Most injection blow mold producers do not make their own secondary injection nozzles, but they purchase them from Viking Co., a small company in New Jersey. Viking also supplies the springs, acorn nut, and block washer for use in the core rods.

I recommend using spherical seated nozzles in any setup. I also recommend that each secondary nozzle have its own heater band. When multiple-cavity injection blow moldings are running, with each secondary nozzle having its own heater band, the operator adjusts the heat to each nozzle individually. If you do not have individual heat controlled secondary nozzles, the only choices are to open the gate diameters for each individual secondary nozzle or to raise or lower the entire manifold heat.

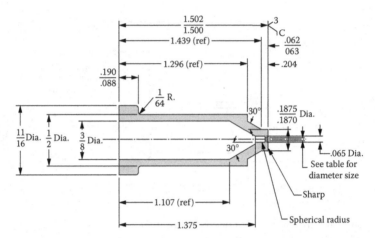

Figure 8.25 Modified secondary nozzle for use with polyvinyl chloride (PVC) and polyethylene terephthlate (PET).

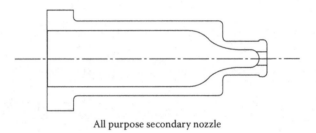

All purpose secondary nozzle

Figure 8.26 All purpose secondary nozzle.

Figure 8.27 provides details for use of a spherical secondary nozzle.

Figure 8.28 is the nozzle guide for use with the spherical secondary nozzle.

There are two (2) main advantages for using a spherical secondary nozzle. With the spherical seat against the parison mold's end cap, there is only a ring contact. This prevents a great amount of heat from transferring from the manifold and secondary nozzle to the end cap on the end of the parison injection mold. The end of the injection molded preform or the gate area is the hottest part of the injection molded preform. The heat coming from the contact of the secondary nozzle adds to the heat of the melted homogeneous resin and is hard to control without the use of the spherical secondary nozzle.

The second advantage is you can design the tooling so each individual secondary nozzle has its own controlled heater band. Thus each secondary nozzle's temperature can be controlled. Any time you are running multiple cavity-molds, if one of the molds is stringing, freezing off, or drooling at the secondary nozzle, and if you don't have individual heat control on the secondary nozzles, you have to raise or lower the entire manifold's temperature

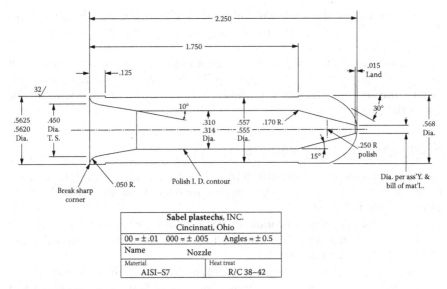

Figure 8.27 Nozzle for spherical seated nozzle.

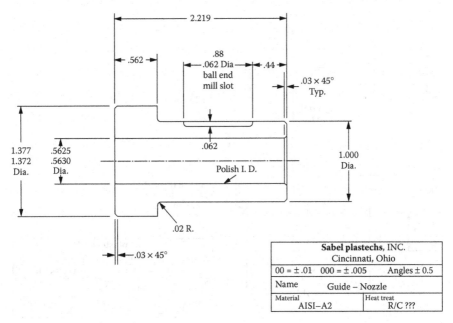

Figure 8.28 Nozzle guide.

to compensate for the problem and this affects all the injection preforms. This defeats the problem of only dealing with the secondary nozzle that is having the problem.

In today's production plants you want to provide the machine operator all the assistance you can. Once again tooling is paid one time, while machine down time and poor quality preforms are continuous and lined with for the life of the project.

Secondary nozzle retainers

All injection blow molding parison injection tooling uses secondary nozzles. Each secondary nozzle has to be held on the heated manifold and these are called nozzle retainers.

Figure 8.29 depicts a secondary nozzle retainer with dimensions.

Figure 8.30 provides the nomenclature with call-outs for specific areas of the secondary nozzle retainer.

Figure 8.31 is another secondary nozzle retainer with dimensions.

Secondary nozzle retainers are designed for each set of injection parison tooling and the particular manifold. They can be relatively standard.

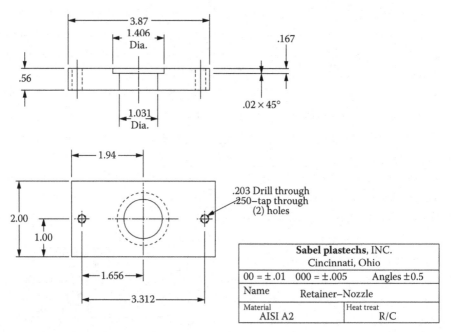

Figure 8.29 Secondary nozzle retainer.

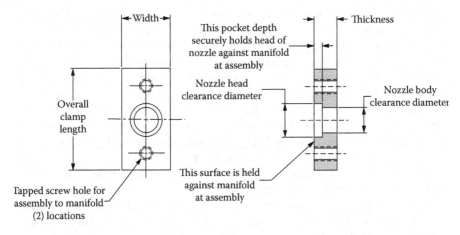

Figure 8.30 Secondary nozzle retainer with proper nomenclature.

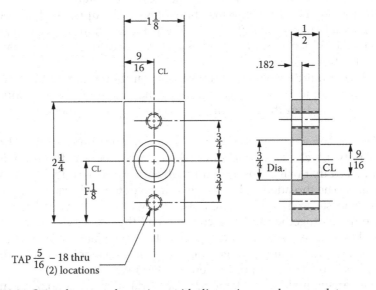

Figure 8.31 Secondary nozzle retainer with dimensions and nomenclature.

Parison mold

The design of the parison mold is the second most critical part of designing for injection blow molding. Naturally, the most critical design is the parison. The parison mold has to produce the parison with as little molded-in stress as possible, take repeated high-injection pressure, withstand constant clamping forces, vent any air before the filling of the cavity, be flash free, and prevent any hot material from sticking to its surfaces.

To determine the number of cavities in the parison mold, you must know the exact length of the trigger bar within the machine in which the parison mold is to be used. The width or maximum diameter of the container to be produced plus 0.5 inch for clearance provides the width; then, by dividing this width into the trigger bar length, you determine the number of center-to-center distances that will fit into this length. If you add one further distance to this number, you have the maximum number of cavities that will physically fit into the specified injection blow molding machine. However, the shot size of the specific machine must be checked to ensure that you have adequate resin for the number of cavities you wish to produce. Normally, in injection blow molding, shot size is not the determining factor.

You must also check the swing radius to ensure that the parisons produced and the blown articles to be produced will clear the manifold and tie bars as the table rotates (see Figure 8.32).

The machine manufacturers will furnish all the dimensions for each specific injection blow molding machine that they produce. They will also furnish a detailed platen layout for the die sets, center of pressure, and clamp tonnage at the parison station and at the blow mold station.

The parison mold has two matching halves. The upper half is mounted to the movable platen, and the lower half is mounted to the fixed platen of the machine. When these two halves are clamped together, you have the true mold shut height.

In laying out the cooling channels, you can figure a minimum of 0.875 inch from center to center. This is a result of the size of the cooling connectors that will be used on the side plates to connect hoses to the thermalator or chiller.

The parison cavity will also have two clearance holes drilled through each cavity block. The 0.5625-inch-diameter holes are for the 0.5-inch threaded rod that will be inserted through each parison cavity block and the end plates. The threaded rod is inserted through the end plates and parison cavities and secured with double nuts. This will secure the end plates and parison molds together and will compress all the "o" rings that have been placed between each parison cavity and the end plates to lock the units together and prevent any leakage from the cooling channels when the tooling is fully assembled and in production (see Figure 8.32).

The parison mold also contains the secondary injection nozzle seat. The seat for the secondary nozzle is critical. This seat has to be of proper fit or heated plastic resin will leak around the secondary nozzle. The secondary nozzle seat provides for the secondary nozzle alignment and proper clearance when the movable mold half closes to the stationary mold half and prevents leakage of the heated plastic resin around the secondary nozzle seat or face. If leakage occurs in this area or any other the injection molded parison will not be adequately filled, causing rejects and eventually the machine will have to be shut down for mold correction.

Figure 8.33 is a side section view of a typical injection parison mold assembly with all the nomenclature.

The parison mold consists of a top and lower injection mold half, a pocket for the neck ring, the water lines and the four (4) holes for the four (4)

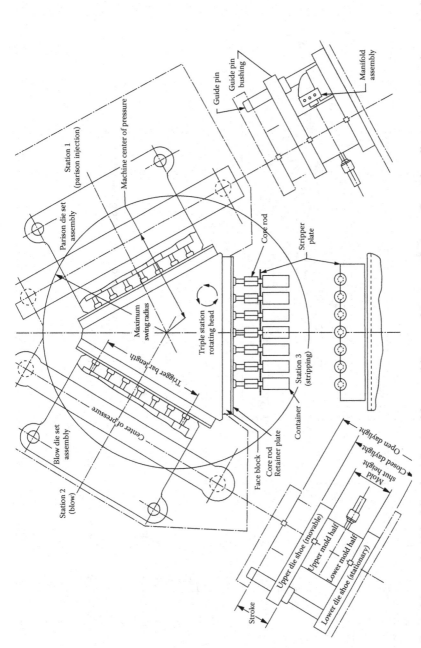

Figure 8.32 Typical layout furnished by the injection blow molding machine manufacturer. Provides all proper nomenclature, including swing radius.

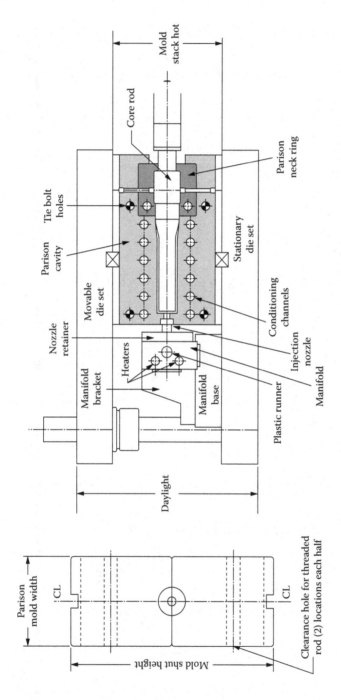

Parison mold setup at the parison station

Figure 8.33 Sectional view of parison mold for injection blow molding.

clamping bolts that hold all the parison molds together. You can see the keyways on the end view of the parison mold and the cross key in the section view of the injection parison mold.

We recommend that a separate end cap be used on the parison injection mold. By going to a separate end cap there are several advantages (see Figure 8.34):

1. In the case of nozzle seat damage, a new end cap can be installed and production continued.
2. A separate cooling channel can be used in the separate end cap, which is critical because the gate area of the parison is the hottest area of the parison. This separated zone provides excellent coolant control to this critical zone.
3. A thermal break can be designed into the bottom injection mold half of the parison. A thermal break is to remove approximately ten thousands (0.010 inch) in depth and one (1) inch wide to each of the parisons mold halfs as they set on the fixed die set. This assists in preventing heat from the heated plastic resin transferring extra heat to the stationary die set.
4. Machining of the parison is easier when there is a separate end cap, as the machining can be accomplished from both ends of the parison block.

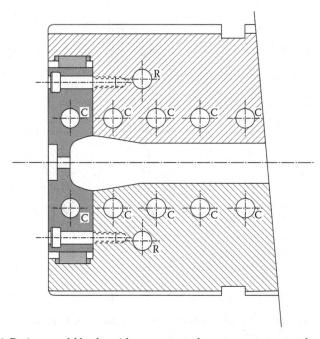

Figure 8.34 Parison mold body with separate end cap to accept a standard secondary nozzle.

5.   When polishing the parison body, it is easier to draw polish, as once again the parison body is open at each end.

Figure 8.34 depicts the use of a standard secondary nozzle in the separate end cap.

Figure 8.35 shows a separate end cap in the parison mold, however the separate end cap is made to accept a spherical seated secondary nozzle.

Figure 8.36 provides all the dimensions for producing the spherical seated secondary nozzle for use with the seperate parison mold end cap.

Figures 8.37 and 8.38 are further details of the separate end cap in the injection parison mold.

We try to design all the parison cavities with a separate end cap. If the secondary nozzle damages this area, a new end cap can be installed. If the secondary nozzle damages the parison mold, the entire parison mold has to be replaced. Costs are always a factor, but production down time is very costly versus the one time charge for the tooling.

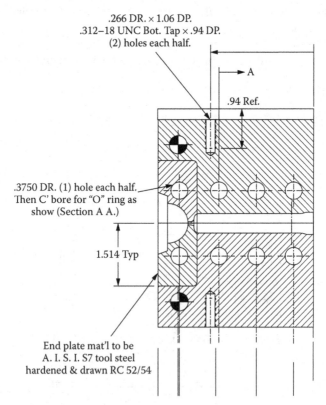

Figure 8.35 Parison mold without neck ring, that displays a separate end cap. For use with a spherical seated secondary nozzle.

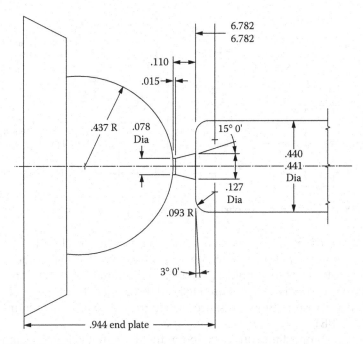

Figure 8.36 Spherical seated end cap details for separate end cap in the parison injection mold.

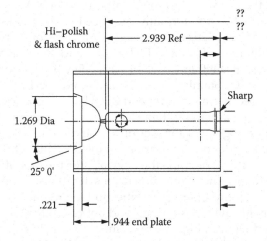

Figure 8.37 Parison mold end cap.

Blow mold

The blow mold setup is very similar to that of the injection parison mold. It consists of two movable halves consisting of a die set, blow mold cavities, side plates, blow mold neck rings, and a bottom plug (split, solid, or movable).

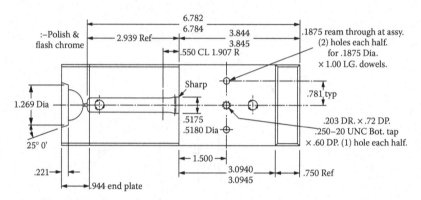

Figure 8.38 Parison setup and parison mold with separate end cap.

Blow molds are used to form the final shape of the article to be produced and are not subject to the same severe pressure conditions as the parison injection mold. The blow molds are subject to minimal mold clamp pressure. The clamp pressure is only required to hold the mold halves together because of the blow pressure. Blow pressures range from 80 psi for the olefins up to 250 psi for PET.

The most important consideration in the blow mold cavity design is the provision for maximum cooling. Once again, as in the parison injection mold, peripheral channeling should be adhered to for maximum cooling.

Chosen on the basis of the type of plastic material to be blow molded, the blow mold cavities are normally produced from aluminum. The types of aluminum used are T-2024, QC-7, and Alumel 89 for olefins, elastomers, ABS, PC, PS, sulfone, nylon, AN, PET Barex, and so on. For PVC to have maximum cooling, use either BECU or Ampcoloy 940 because of the corrosive activity of PVC. Stainless steel is also used for PVC; however, you sacrifice cooling because aluminum and BECU have approximately seven to eight times better conductivity than steel.

The die set arrangement for mounting the die set in the injection blow molding machine is identical to the parison injection die set. The guideposts, die set thickness, and so on, plus the keyways for the mounting of the blow mold cavities, follow the same dimensions for the keyways and for the mounting of the injection parison cavities to the die set.

The blow mold cavity, neck rings, and bottom plug yield a product that matches the product drawing. The blow mold tooling has shrinkage cut into all the dimensions in the metal. This shrinkage is based on a product history file kept by the injection blow molding processor and on the resin data obtained from the resin supplier. Normally, when producing a high density polyethylene (HDPE), the shrinkage factor will be 0.018 inch/inch up to 0.022 inch/inch on the diameters and 0.012 inch/inch up to 0.015 inch/inch. This means that for each inch dimension of the desired blow molded bottle the blow mold must be built large enough so when the bottle shrinks upon cooling, it will have the dimensions specified by the bottle user. This is achieved by multiplying the desired bottles diameter by 0.018 inches or by 0.022 inches to have the blown bottle shrink upon cooling to be within the bottle customers specifications. If the bottles diameter is to be 3.00 inches, then multiply 3.00 times 1.018 to have the mold diameter to be 3.054 inches. This is also done for the height. Plastic bottles do not have the same shrinkage in the diameter as they do in the height. Injection blow mold shops know the project shrinkages to use based on their extensive years of producing injection blow mold tooling. Shrinkage dimensions can be fine tuned on the tooling when a unit cavity is used. Normally, a unit cavity will be built before production tooling is ever built. It is with this unit cavity that parison design, venting, furnish dimensions, etc. are all proven.

Figure 8.39 provides detail for venting one half of the blow mold. Only the top or bottom half is vented.

Figure 8.40 is a side view of a typical injection blow mold set-up. You can see the bottle outline, the water lines, the four (4) holes for the retaining bolts, the blow mold neck ring, the keyways, the split bottom plug or push-up, the core rod, and the die sets.

The blow mold will normally be produced using a quality grade aluminium such as QC-7 and ALUMEC 89. The push-up or bottom plug may be aluminum or amcoloy. The neck ring will be steel.

Parison and blow mold neck ring

The parison neck ring forms the finished shape of the threaded or neck section of the container. It also centers and securely retains the core rod inside the injection parison cavity to aid in preventing core rod deflection during the injection phase of the total blow molding cycle.

The parison neck ring details are determined by the finish, which is specified by the customer (i.e., 28-A400), and by multiplying the desired

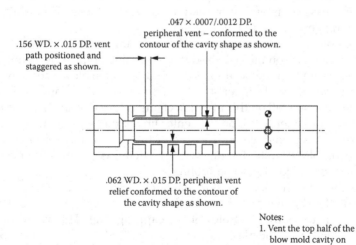

.047 × .0007/.0012 DP.
peripheral vent – conformed to the
contour of the cavity shape as shown.

.156 WD. × .015 DP. vent
path positioned and
staggered as shown.

.062 WD. × .015 DP. peripheral vent
relief conformed to the contour of
the cavity shape as shown.

Notes:
1. Vent the top half of the
blow mold cavity on

Figure 8.39 Normal venting of a blow mold for high density polyethylene.

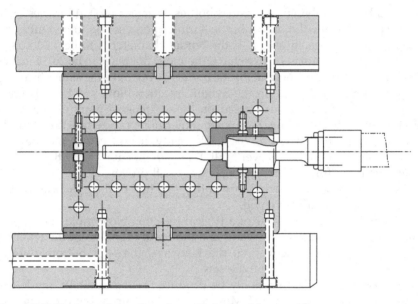

Figure 8.40 Typical side section view of injection blow mold.

dimensions by the appropriate shrinkage factor for the specified plastic resin to be used for the product. The parison neck ring is most commonly produced from non-deforming hardened A-2 tool steel.

Figure 8.41 provides detail and nomenclature for the parison mold and shows the pocket for the neck ring insert.

Figures 8.42 and 8.43 provide details for the neck ring insert for the parison mold.

The injection parison cavity also has hold-down screw holes to hold the neck ring halves in position. Each parison neck ring should be marked upper or lower for ease of assembly of the parison cavity block. This is also true of the separate end caps and side plates.

Please note the center line of the injection clamp is the center of pressure for the injection parison molding and the center line of the blow clamp is the center of pressure for the injection blow molding of the blown container. The injection parison mold designer should strive to have the center of the injection molded parison fall on this injection clamp center line and also the center of the blown container to fall on the center line of the blow clamp centerline.

Figure 8.44 is a engineered drawing of the thread finish desired on the final injection blow molded container. This detail is cut into the neck ring for the injection molded parison tooling.

Note the "T" diameter is the diameter across the threads. The "E" diameter is the diameter of the neck finish less the height of the threads. The "B" diameter is the diameter of the capping ring. This is old nomenclature retained from the glass bottle industry. "TS" stands for True Sharp. BLD stands for blend i.e. a 0.005 Radius blend. The T, E, B, are all Standard

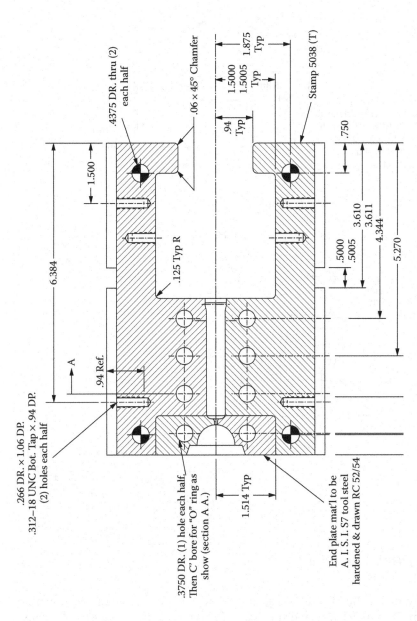

Figure 8.41 Nomenclature and dimensions for parison mold depicting parison body water lines and separate end cap and neck ring pocket.

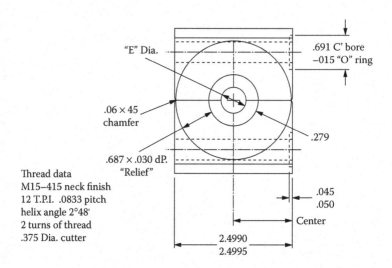

Figure 8.42 Parison neck ring details.

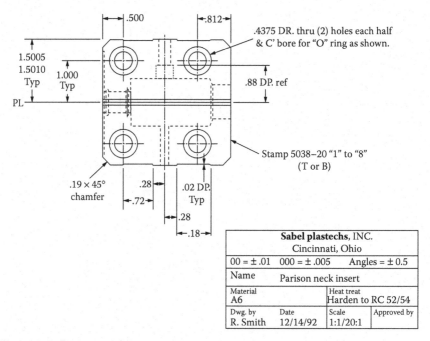

Figure 8.43 Parison neck insert.

Society of Plastics Industry callouts for each specific size neck finish i.e. 33 mm, 28 mm etc. all have specific T, E, H, I dimensions in the bottle blow molding industry.

The actual thread finish is shown in Figure 8.45. Each thread finish has its own clearances and dimensions.

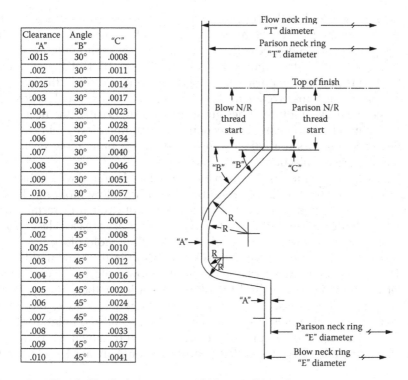

Clearance "A"	Angle "B"	"C"
.0015	30°	.0008
.002	30°	.0011
.0025	30°	.0014
.003	30°	.0017
.004	30°	.0023
.005	30°	.0028
.006	30°	.0034
.007	30°	.0040
.008	30°	.0046
.009	30°	.0051
.010	30°	.0057
.0015	45°	.0006
.002	45°	.0008
.0025	45°	.0010
.003	45°	.0012
.004	45°	.0016
.005	45°	.0020
.006	45°	.0024
.007	45°	.0028
.008	45°	.0033
.009	45°	.0037
.010	45°	.0041

Figure 8.44 Finish details for parison and blow mold neck ring.

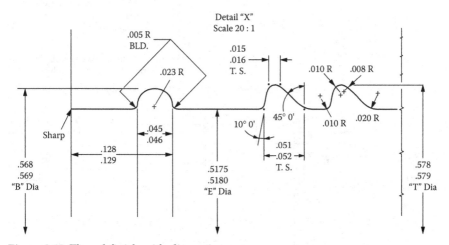

Figure 8.45 Thread finish with dimensions.

The clearance cannot be excessive; otherwise, when air enters to lift the injected parison from the core rod, the finish can be partially blown. This condition can cause an irregular blown product, as the blow air can escape around the core rod and the finish fall outside the desired specifications. The clearance normally is between 0.001 and 0.005 inch per side.

In all blow mold neck rings, the start of any thread is always located 5 degrees before the start of the thread, as cut into the parison injection mold neck ring, and ends 5 degrees after. The 5 degrees of separation are shown as the thread cutter start and end positions. This is part of the clearance needed to ensure that the finish (threads) is not damaged when the injection molded parison is transferred to the blow station via the core rod. It is possible to change the neck ring thread design or neck finish, because the neck rings are inserts into both the parison injection cavity andthe blow mold cavity.

Figure 8.44 shows normal clearances used in the blow mold neck ring, versus the parison injection cavity dimensions.

Figure 8.46 provides details for a blow mold neck ring in which the neck ring insert is round, fitting to a round pocket in the blow mold block. The cost of machining in a mold shop can be a factor, of circular or rectangular; however, we prefer the rectangular to have more surface area and less chance of movement in either the parison block or the blow mold block.

Figure 8.47 is for a rectangular neck ring insert in the blow mold.

There are usually two pockets cut into the blow mold cavity block. One pocket is for the neck ring, and the other pocket is for the stationary split bottom plug or the movable bottom plug. The blow mold cavity block will also have cooling channels, hold-down screw holes, and rod holes, for holding the end plates and compressing the o-rings, all of which are patterned after the parison injection mold. The processor should always be set up with

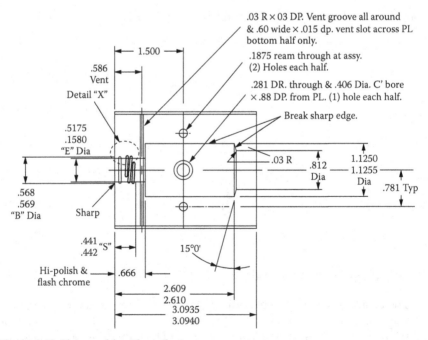

Figure 8.46 Blow mold with neck ring.

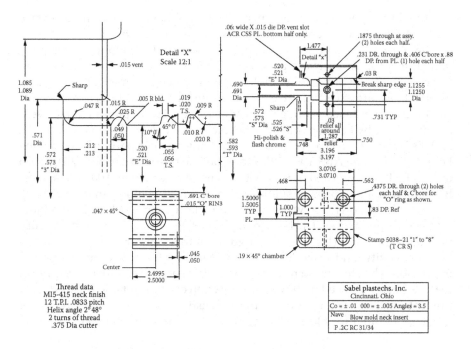

Figure 8.47 Rectangular blow mold neck ring with details.

a separate cooling zone on the neck rings in both the parison injection mold and the blow mold cavity. The bottom plug may or may not have a cooling channel. Normally, movable bottom plugs have their own cooling circuit. Whether or not to have cooling channels in the bottom plug is usually choice of the processor and the mold designer. With the slow-setting resins, such as the olefins that are slow giving up their heat, with PET to prevent gate crystallization, and with high–melt temperature resins, it is best to design a separate cooling channel for the bottom plug.

The blow mold cavity body should be high polished and chromed for materials as PET, AN, Barex, ABS, and nylon PS, as well as sulfone and PVC in some instances. When running the olefins such as HDPE, LDPE elastomers, and so on, the blow mold cavity body is vapor honed, sand blasted, or glass bedded, depending on the customer's surface needs. Normally, a 180 grit is specified. However, this surface finish has to be the customer's choice.

Venting is always a concern in any blow molding process. There are face vents, slot vents, pin vents, segment vents, and surface finish venting used on the blow mold body cavity. Naturally, where the bottom plug is positioned, there is an excellent area to vent the area around the bottom plug insert. If the body of the blow mold cavity has been segmented or contains inserts, these are excellent areas to have semicircular venting.

I recommend face venting in most instances. The depth of the face vent is determined by the resin that is being blow molded. For olefins, a face vent of 0.0005–0.0012 inch can be used without any flash or parting line display. For products using ABS, PVC, PS, and so on, the face vent can be 0.001–0.0025 inch. If PET is being used, it is possible to use a face vent of 0.004–0.006 inch. Elastomers usually behave similar to an olefin in venting. In each design, there are hold-down screws for securing the neck ring block in the blow mold neck ring pocket. The blow mold neck ring provides core rod support, maintains the finish (threaded) configuration that was formed in the parison injection cavity, and provides a shut off for the core rod when it is seated in position. The blow mold neck ring has "clearance" to ensure that the finish that has been produced in the parison injection cavity is retained. This should prevent pinching or any marring to the finish as the injected parison drops in the blow mold for blow molding the injected molded parison.

Bottom plug (push-up)

The bottom plug in the blow mold cavity forms the bottom configuration of the final article. It is usually produced in two (2) halves, with each half contained in each half of the blow mold cavity blocks. It may be in one piece and can have its own cooling circuit. In this instance, it would be part of the bottom or stationary section of the blow mold tooling.

For the olefins, push-up heights up to 1/16 inch are possible without a movable bottom plug, and the article is stripped from the lower half of the blow mold cavity when the core rod containing the blown article is raised out of the down position on the blow mold lower half cavity block.

With rigid materials as polystyrene, polycarbonate, polysulfone, nylon, Barex, acrylonitrile, and some of the new metallocenes, a maximum push-up height cannot exceed 1/32 inch. For the rigid materials that require greater than a 1/32 inch push-up height, a retractable bottom mechanism is necessary.

By using retractable bottom plugs, it is possible to manufacture rigid products with up to a 3/8 inch push-up height. This design requires an air cyclinder, cam, or spring actuated mechanism which retracts the one-piece bottom plug prior to, or simultaneously with, the blow mold opening. The latter requires careful design, especially when the maximum depth push-up is required due to the simultaneous movement of the bottom plug and the finished product.

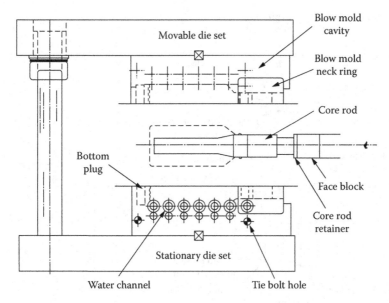

Figure 8.48 Typical injection blow mold with two (2) piece bottom plug.

Figure 8.48 shows an ordinary injection blow mold with all the nomenclature. Notice the bottom plug call out. It is a two piece design used for the olefins and the depth of the push-up is less than 1/16 inch.

Due to costs and the mold designers choice, they may decide to use a one piece bottom plug. Again, the design has height limitations and also what resin is being used to produce the injection blow molded product. Figure 8.49 is the normal one piece push-up.

The depth of the push-up specified by the end product user will usually determine if one (1) of the two (2) bottom plugs or push-ups may be used.

If the depth of the push-up is greater than 1/16 inch or the resin to be used is of the rigid type as polystyrene, polyethylene terephthlate, poly carbonate, cyclic olefin copolymer, etc., the mold designer may choose to use a cammed bottom plug. Figure 8.50 shows the design of a cammed bottom plug or push-up.

Figure 8.51 provides the mold designer with details for designing the cammed bottom plug or push-up.

The disadvantage of any of the push-ups described is the problem of not having a water circuit in the push-up to remove heat from the push-up area of the injection blow molded product. Keep in mind the hottest part of the injection molded parison is the gate area and this area will be in the center of the injection blow molded bottom or push-up area. If you are running polyethylene terephthlate (PET) the gate area may crystallize if this gate area is not cooled. This will cause a small white area to show in the bottom of the bottle. This area can be as large as a 1/4 inch in diameter and will cause the product blown to be rejected.

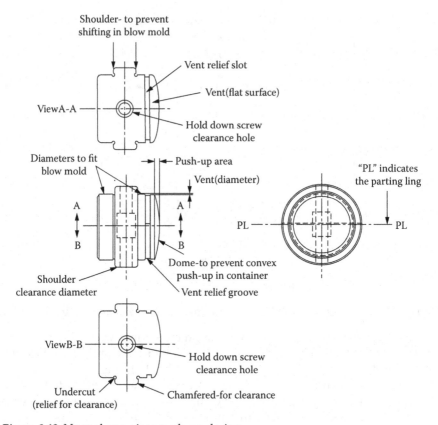

Figure 8.49 Normal one piece push-up design.

If drop impact is a problem on small containers, either the neck radius or the push-up will have to be modified to provide cushioning for drop impact or a cooling circuit will be needed to reduce the residual stresses that are molded in the parison and present when the injection molded parison is blown into the bottle.

I have never seen a mold of any type that I had too many cooling circuits, but I have seen many molds where I wish I had a cooling circuit. If there are extra cooling circuits you can always throttle the water flow, but you can't drift a new water line. Thus, when in doubt, the cooling circuit should be designed in the bottom plug or push-up and utilized.

The final push-up or bottom plug is the use of a roller ball with a cam that is spring loaded to retract the bottom plug or push-up.

The roller push-up design is used in the injection blow molding industry. It is the most expensive push-up or bottom plug design.

Figure 8.52 is a depiction of the actual push-up used in the roller cam design.

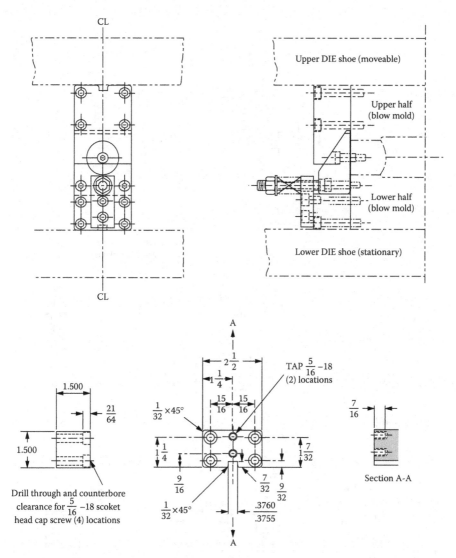

Figure 8.50 Typical cammed bottom plug design.

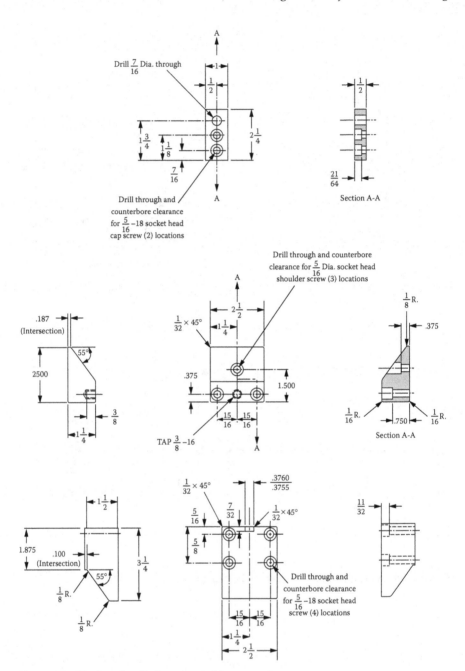

Figure 8.51 Design details for cammed bottom plug.

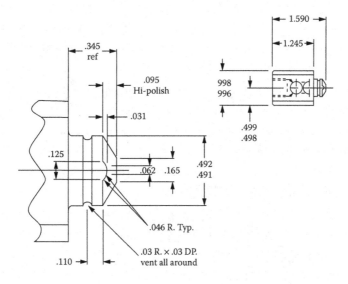

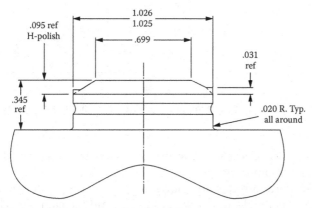

Figure 8.52 Actual push-up used in the roller cam design.

Figure 8.53 shows the bottom plug dimensions for a flat oval injection blow molded bottle design used with the roller cam design.

Figure 8.54 shows the bottom plug design for use with an oval injection blow molded bottle and used with a roller cam arrangement.

Figure 8.55 is the push-up or bottom plug design when producing a round injection blow molded bottle using the roller cam design.

Figure 8.56 depicts the components for the roller cam push-up or bottom plug.

The amount of the push-up travel is determined by the roller ball contact with the cam angle.

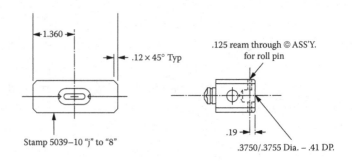

Stamp 5039–10 "j" to "8"

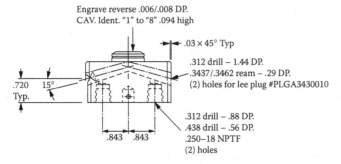

Sabel plastechs, INC. Cincinnati, Ohio		
00 = ± .01 000 = ±.005		Angles ±0.5
Name	Bottom plug	
Material Alum. OC7		Heat treat

Figure 8.53 Bottom plug with dimensions for flat oval container with roller cam design.

Figure 8.54, 8.55, and 8.56 show the components used with the roller cam push-up or bottom plug.

Housing for bottom plugs are shown in Figures 8.57, 8.58 and 8.59.

The final part is the design of the guide ball and the cam bar. These are shown in Figure 8.59.

After reviewing the four push-up or bottom plug designs, you can realize each design has advantages and disadvantages. Naturally, the roller cam design is the most expensive to produce. If you must have cooling water in the push-up or bottom plug then the roller cam design is the choice.

Side plates

The side plates used on the blow mold can be identical to the side plates used on the parison injection cavities. They may also be different. They will always have two (2) drilled through 7/16 inch diameter holes to allow for threaded rod to pass through the side plates and all the blow mold blocks, to pull all the blow mold cavities together to compress the "o" rings between

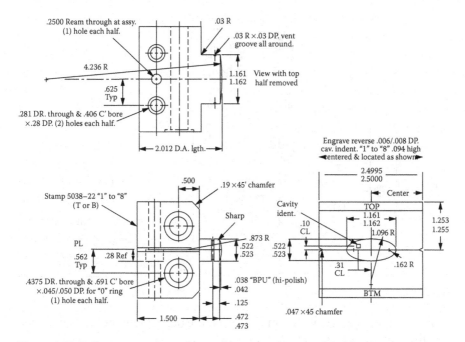

Figure 8.54 Roller cam push-up for oval injection blow molded bottle.

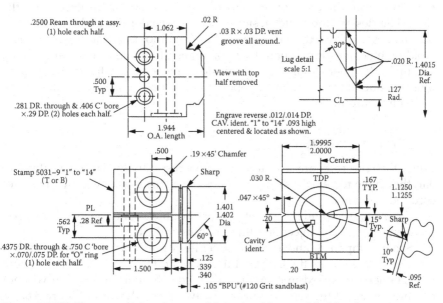

Figure 8.55 Roller cam design for round injection blow molded bottle.

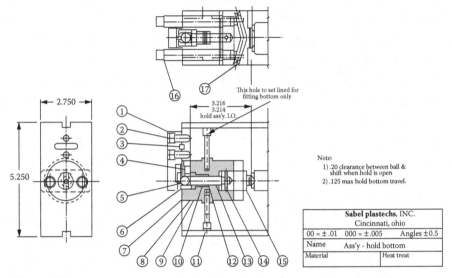

Figure 8.56

each blow mold cavity blocks and between the outer blow mold cavity block and the matching side plate. See Figure 8.60.

The number of water lines on the side plates is determined by the length of the parison and the blown product. You should always have a water circuit under the neck rings, the end cap and the area where the neck ring transitions into the body of the parison. In doing the mold design and the side plate design, the mold designer has to consider the mold set-up with water lines to allow the person installing the molds in the injection blow molding machine ample room to use a wrench to tighten each water filling. The mold designer should always consider the person installing the tooling plus the operator to enable the operator to produce quality product from the tooling supplied.

Core rod retainers

When the core rods are inserted into their respective positions in the face bar, they have to be held securely in position. This is achieved by the use of core rod retainers.

Figure 8.61 provides the details for the most common used core rod retainers used in the injection blow molding industry.

Some companies and machine operators prefer a two piece core rod retainer. This set runs the length of the face bar. I prefer the individual core rod retainers for speed in changing core rods while in production without having to take all the screens out on the two piece shown in Figure 8.62.

Anyone using injection blow mold tooling should consider mold change over time, ease of set-up, and the operator.

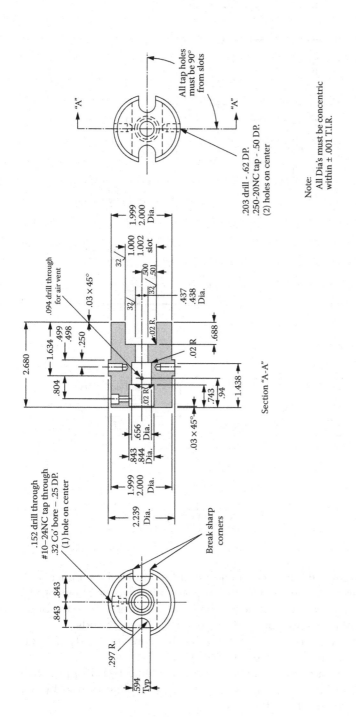

Figure 8.57

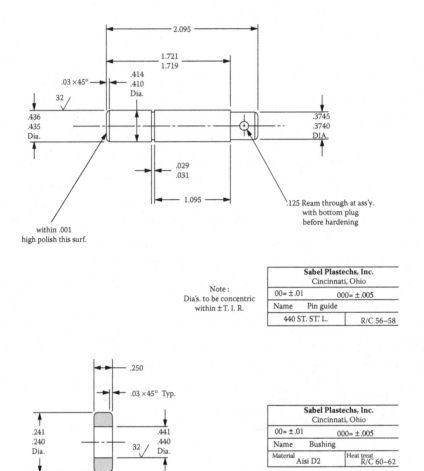

Figure 8.58 Pin and bushing for use in roller cam design.

Stripper

Schematic A in Figure 8.63 depicts an injection blow molded container on a core rod ready to drop into position for the stripper to remove the blown bottle.

Each injection blow molding machine has a strip station where the blow molded plastic containers are stripped or pulled off the core rods.

Schematic A shows the blown container on the core rod ready to drop into the stripper to be removed from the core rod.

Schematic B in Figure 8.64 shows the rotating table has dropped into position placing the blown container in the stripper plate so the blown container can be removed from the core rod.

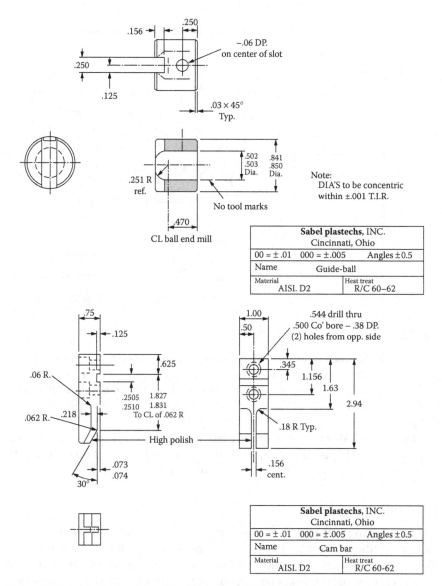

Figure 8.59 Guide ball and bushing for use with the roller cam assembly.

Schematic C in Figure 8.65 shows the blow bottle being ejected with the stripper moving out and once the bottle is dropped into a container or on a conveyor belt, the stripper returns for the next cycle.

On most machines the stripper is able to rotate 90° downward to deposit the product onto a conveyor, etc. We refer to this stripper as a stripper/tipper.

The stripper consists of a stripper base that is part of the injection blow molding machine. The base will have an air line connection plus small 0.040

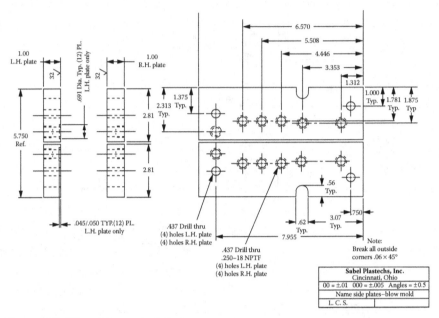

Figure 8.60 Side plates typical design.

inch diameter holes along the air passage way for use in cooling the tip of the core rods, if the operator desires. Figure 8.66 depicts the air inlet and the outlets plus holes for mounting the stripper plate. The actual length shown of the stripper base is for a specific size injection blow molding machine design.

Figure 8.67 shows the stripper plate for a seven (7) cavity specific size container. The stripper plate is part of the mold design and is furnished with all the tooling. It is usually produced from a quality grade aluminum.

Face bar

The face bar mounts to the rotating table and its function is to hold the core rods in position. The face bar has to be designed for the specific injection blow molding machine rotating table and for the number of core rods that are going to be used in each particular set-up. They are part of each tooling package.

Figures 8.68 and 8.69 provide details of a face bar. This is for a seven (7) cavity set-up.

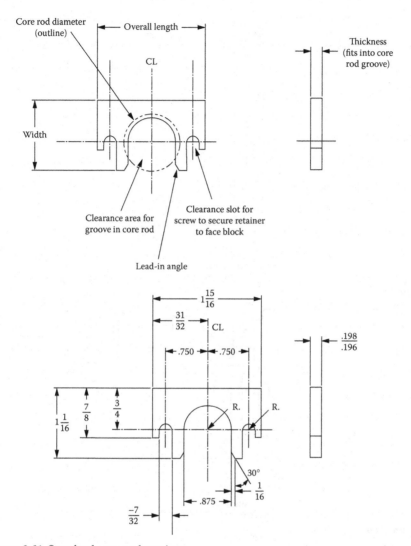

Figure 8.61 Standard core rod retainer.

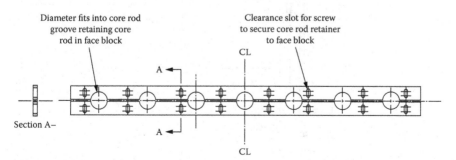

Figure 8.62 Two piece core rod retainer.

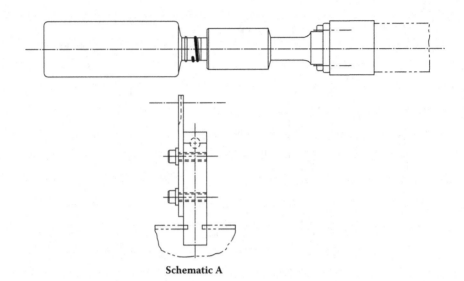

Schematic A

Figure 8.63

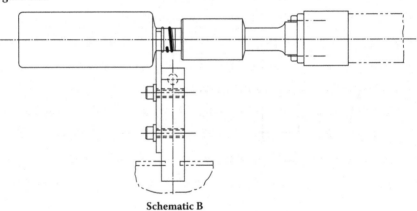

Schematic B

Figure 8.64

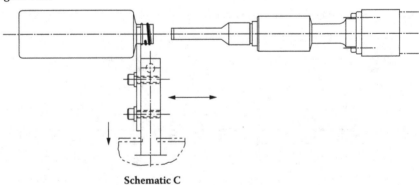

Schematic C

Figure 8.65

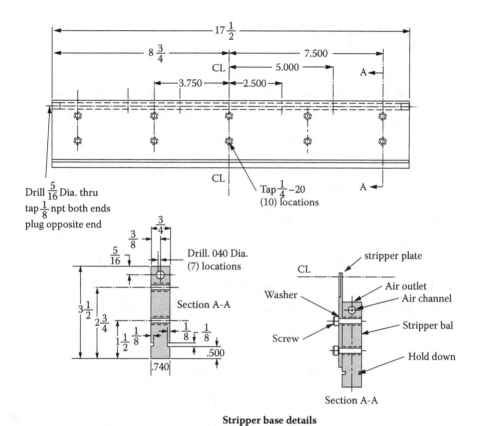

Stripper base details

Figure 8.66

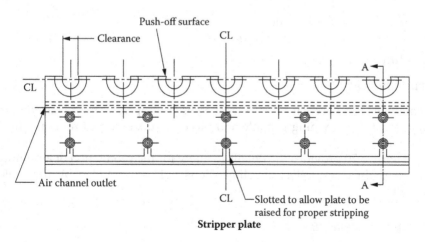

Stripper plate

Figure 8.67

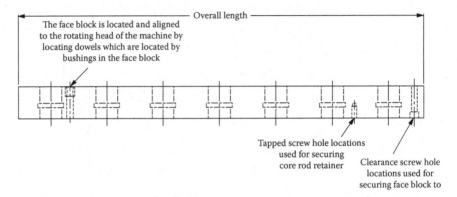

Figure 8.68 Face bar nomenclature.

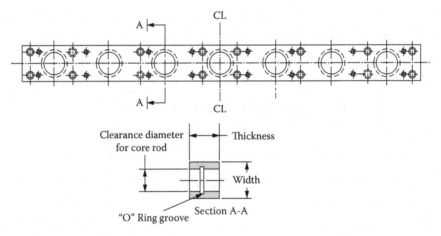

Figure 8.69 Typical face bar design.

Extras

Dummy plug

Sometimes it is necessary to block off a core rod opening in the face bar. In this case a dummy plug is made and placed in the hole that normally held a core rod.

Figure 8.70 provides the details to allow for the production of a dummy face bar plug. Naturally, the dimensions will be provided by the mold designer.

Dummy secondary nozzle

It is possible that once in a while, it is necessary to block the secondary nozzle exit from the manifold. In this case a dummy secondary plug is produced and placed on the manifold. It has the outside dimensions of the

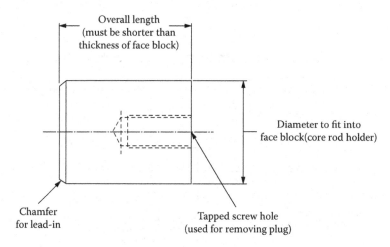

Figure 8.70 Dummy face bar plug.

secondary nozzle being used with the exception that it has no opening for the heated resin to leave the manifold and enter into the injection preform mold.

My recommendation is to use neither of these tools.

Figure 8.71 provides detail for the dummy secondary nozzle.

Each injection mold tool designer has their own ideas, companies they purchase tooling from, steel companies that supply their tool steels and what capabilities the mold shop has that will build the molds and tooling. The table is to used as a guide.

Materials for tooling

Table 8.2 is provided to allow for the reader to have a guide for use when designing injection blow mold tooling.

Each injection mold tool designer has their own ideas, companies they purchase tooling from, steel companies that supply their tool steels, and what capabilities the mold shop has that will build the molds and tooling. The table is to be used as a guide.

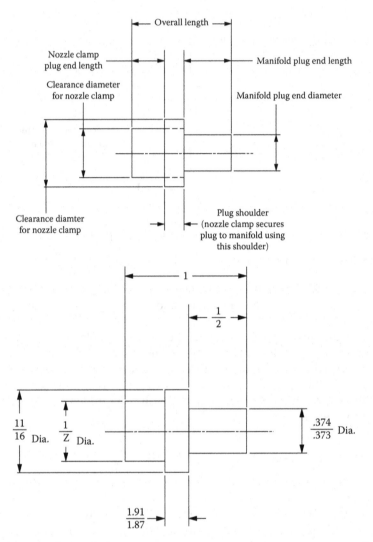

Figure 8.71 Dummy secondary nozzle.

Table 8.1

Part	Figure
Manifold	Figure 8.12
Manifold heaters	Figure 8.12
Manifold insulation	Figures 8.18–8.20
Secondary nozzle retainers	Figures 8.29–8.31
Secondary nozzles	Figures 8.23–8.26
Bill of Materials	Figure 8.22
Sprue Bushing	Figure 8.17
Gusset-manifold	Figure 8.14
Mounting plate	Figure 8.13

Table 8.2 Materials for Tooling

Item	Material	Recommended Hardness	Recommended Surface Finish
Die sets	1020 H.R.S.	Stress relieved	Nickel plate
Die sets	AA-1623-70	Stress relieved	Nickel plate
Manifold soft material	1020 H.R.S.	N/A	Nickel plate outside
Manifold soft material	DME #1 steel	N/A	Smooth polish all runners and chrome
Manifold (PVC)	420 stainless		Runner SPI #1 polish
Manifold base	1018 C.R.S.	N/A	N/A
Manifold spacer	1020 H.R.S.	N/A	N/A
Manifold support	1020 H.R.S.	N/A	N/A
Secondary nozzle (PVC)	AISI H-13	Rc 48–50	Polish and chrome
Secondary nozzle (PVC)	420 Stainless		Polish runner
Nozzle guide	Latrobe Viscount	Rc 44	Polish and chrome
Secondary nozzle (soft resin)	AISI L-6	52–54 Rc	Polish and chrome
Secondary nozzle (soft resin)	AISI P-20 tool set		Plate runner
Nozzle retainer	AISI S-7	Rc 48–50	N/A
Manifold nozzle	1018 C.R.S.	N/A	N/A
End cap parison mold	1018 C.R.S. NAK 55	N/A	Polish and chrome
End cap blow mold	1018 C.R.S.	N/A	Polish and chrome
Parison mold	1018 C.R.S. NAK 55	N/A	Polish to SPI #A1 Chrome plate Cavity nickel plate H_2O lines
Parison mold	420 stainless		Polish and chrome Plate cavity
Parison neck ring	AISI A-2 tool steel	52–54 Rc	Polish and chrome Plate–nickel plate H_2O lines
Core rod body	AISI A-6 tool steel	52–56 Rc	Draw polish Chrome plate
Core rod tip	AISI A-6 tool steel	52–56 Rc	Polish and chrome Plate
Core rod acorn nut	1117 C.R.S.	Case harden	N/A
Core rod lock nut	1018 C.R.S.	Case harden	N/A

(Continued)

Table 8.2 Materials for Tooling (*Continued*)

Item	Material	Recommended Hardness	Recommended Surface Finish
Core rod spring	.051 Music Wire	N/A	N/A
Threaded rod	Steel	N/A	N/A
Blow mold cavity	AISI A-2	52–54 Rc	Polish and chrome Plate–nickel plate H_2O lines
Blow mold cavity	QC-7 AL	N/A	Polish and chrome Plate
Blow mold neck ring	QC-7 AL	N/A	Olefins (Sand blast) PVC (Chrome)
Blow mold neck ring	AISI Tool Steel	52–54 Rc	Polish and chrome Plate
Bottom plug	QC-7	N/A	Polish
Bottom plug	Ampcoloy 940	N/A	Polish
Bottom plug	AISI A-2 Tool Steel	45–48 Rc	Polish and chrome Plate
Wedge block	H-13	55–56 Rc	
Support block	101T C.R.S.	N/A	N/A
Spring	Steel	N/A	N/A
Mounting block	AISI A-2 Tool Steel	52–54 Rc	N/A
Stripper plate	C.R.S. OR AL	N/A	Nickel plate C.R.S.
Stripper base	C.R.S. OR AL	N/A	Nickel plate C.R.S.
Face bar	1020 H.R.S. Aluminum QC-7 Aluminel 89	N/A	N/A
Core rod retainer	AISI O-1 tool steel	50–52 Rc	N/A
Face bar plug	QC-7/AL 7075-T6	N/A	N/A
Nozzle heater	Watlow coil heater	0.125-inch R_D × 0.500 in I.D . × 1.00 inch long (15 watts, 120 volts) leads SS hose 6 feet long	
Side plates	Flat Gr. Stock	N/A	Nickel plate entire surfaces
Parison gate insert	AISI A2	Rc 52–54	SPI #2 finish and chrome

chapter nine

Mold setup and installation

It is always wise to check out the machine to be used before installing any tooling. Checks that should be completed by the person ready to install tooling are as follows:

1. Check oil reservoir
2. Check lubricating oil (Trabon or Bijur)
3. Check air pressure
4. Check stripper
5. Turn all selector switches to off on the operators' panel
6. Set the automatic/manual selector to manual
7. Check all safeties

Once in the manual mode and the hydraulics are turned on, check the stripper, clamp (open-close) table index, manual head cylinder, plasticizer positioning cylinder, tip cooling, and head (on/off) switch. Once the heats are turned on and up to temperature, the screw rotation, injection and decompression should be checked.

The machine and the surrounding area should be clean and free of any debris.

The following page representations are taken from the pages of the Uniloy Milacron Rainville mold setup instructions, mold setup, and tool list (courtesy of Uniloy Milacron Rainville) that outline the mold setup procedures.

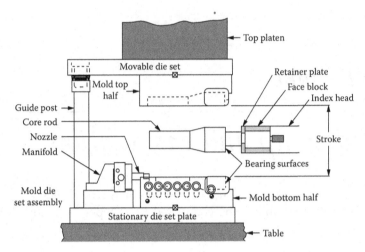

Figure 9.1

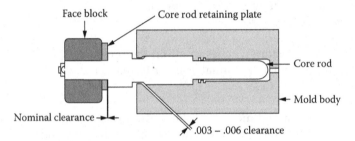

Figure 9.2 Proper fit of core rod.

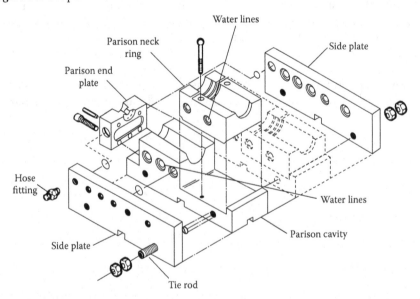

Figure 9.3 Parison mold assembly.

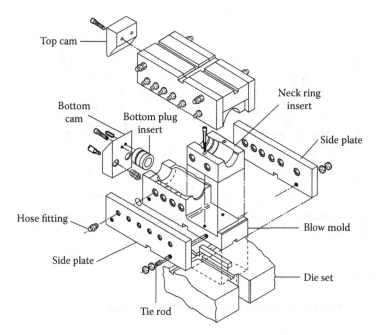

Figure 9.4 Blow mold assembly.

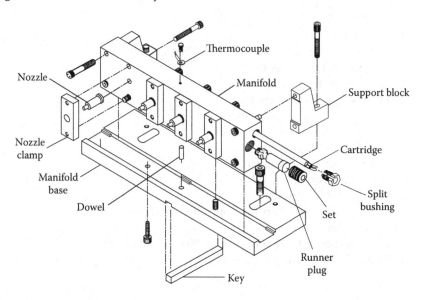

Figure 9.5 Manifold assembly.

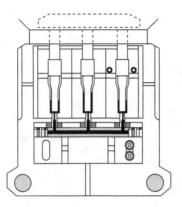 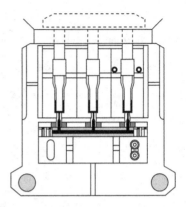

Figure 9.6 Unbalanced material flow from manifold.

Figure 9.7 Balanced material flow from manifold.

Mold setup tool list

Tools you should have for changing molds.

> 1) 5/32" Allen Wrench.
> 1) 3/16" Allen Wrench.
> 1) 5/16" Allen Wrench
> 1) 3/8" Allen Wrench
> 2) 5/8" Allen Wrench. (1 Long and 1 Short)
> 1) 5/8" Open End Wrench
> 1) 3/4" Open End Wrench
> 1) 11/6" Open End Wrench
> 1) 5/8 × 3/8" Drive, Deep Socket
> 1) 11/16 × 3/8" Drive, Deep Socket
> 1) 3/8" Ratchet Drive
> 1) 3/8 × 3" Ratchet Extension
> 1) 1/4" Allen Key Socket 3/8" Drive
> 1) 6" Adjustable Wrench
> 1) Special, Core Rod Wrench (Supplied With Machine.)
> 1) 12" Pry Bar
> 1) 1" Micrometer. (.001" Grad.) For Measuring Shims
> 1) Roll 1/2" Wide Teflon Pipe Tape

Mold setup instructions

Stop!! Read Instructions Carefully Before Proceeding

1. Turn on the cooling water to the heat exchanger.
2. Turn on external main power disconnect.

3. Check that the key switch on operator's panel is in the "manual", "Off" position.
4. Turn on the main circuit breaker or the front of the electrical control panel.
5. Check to assure that all selector switches on the operator's panel are in the "manual", "Off" or "Up" position.
6. Turn the key switch "On." Reset light will come on.
7. Push the "Reset" push button, system will reset. (SCI System)
8. Turn the pump motor switch momentarily to "Start"; and when you hear it start, release it to the "Run" position.
9. Turn clamp switch to "Open." The clamp will open and an amber "Auto Start" light will come on.
10. Open safety gate if not already open.
11. Make sure mechanical safeties are in position against tie-bars, and that they are under the cross arm lock nuts on the injection and blow mold stations.
12. Now turn the pump motor switch to the "Off" position.
13. Stone and clean the table and under the upper platens to remove any burrs and dirt. Make sure no dirt or debris remain.
14. Coat the table lightly with thin oil.
15. A) Put the parison (injection) mold die-set (Figure 9.1) on a die cart and raise to a convenient working height. Clean the bottom of the die-set and install hose fittings in all ports. Now move the die-set to the back side of the machine table and raise the die-set to about 1/4" above the level of the table. Make sure the core rod opening holes are facing toward the three-sided indexing head when the die-set is slid onto the machine table.

    B) Push the die-set toward the index head until it engages the locating key. Push forward toward the index head until it is almost bottomed out against the spacer plate.

    C) Now place .050" shims on both sides of the key, between spacer plate and die shoe, about halfway between key and the end of the die-set. (On larger machines put the shims as far out as spacer plate allows.)

    D) Now push die-set so the mold is against the shims until they are both tight. This insures that the mold is parallel with the indexing head.

    E) Clamp the bottom half of the die-set to the table. Recheck that the shims stayed tight.

16. Repeat the step 15 for the blow mold **except use .055" thick shims (.005") thicker.** Use only .050" thick shims when processing rigid material like Styren, Barex, PVC, etc. (This change of shims is for alignment of threads in the blow station that were molded at the injection station, as there is a growth or shrinkage change during transfer.)

17. After both die-sets have been installed and clamped to the table, clean the top surface of each die-set. Stone, if there are any burrs.
18. Turn pump motor switch to "**Start**" position momentarily. Pump motor will start.
19. Turn clamp switch to "**Open**" to insure that the clamp is in the full open position. It may have sagged down onto the safety arms while the pump was turned off.
20. **Stop here** and pull down the safety gates, making sure that the safety arms move out of the way of tie bars and cross arms.
21. Turn the clamp selector switch to "**Close.**" Clamp should close.
22. Turn the pump motor switch to the "**Off**" position.
23. Open safety gate.
24. Install the top bolts, to attach the movable plate of the die-set to the top platen and tighten. Make sure the bottom plate of the die-set is clamped to the table.
25. Close both safety gates.
26. Turn pump motor switch to "**Start**" momentarily. Pump motor will start.
27. Turn the clamp switch to "**Open.**" Clamp will open. Hold in this position until the clamp stops moving up. It is now in full up position.
28. Remove the core rod retaining plates from the face blocks. Set aside all parts.
29. A) Install one (1) face block (core rod holder Figure 9.1) to one of the three stations of the indexing head. Make sure the 1/4" dia. dowels are in the head for alignment of the face block. Also make sure that the trigger bar is in the full back position.
    B) If the face block is tight when fitting to the dowel pins, hit the face block with the palm of your hand or a soft hammer until it seats against the indexing head.
    C) Install all the #10-32 cap screws used to hold the face block to the head. Do not leave any of these screws loose, or they may come out and fall into the mold, causing serious damage.
30. A) Install a core rod in each extreme outboard cavity.
    B) Be sure that the slot for retaining plate, on the core rod, is flush with the front surface of the face block. (Refer to Figure 9.2)
    C) Install the retaining plate and tighten the bolts.
31. Index this station around to the injection mold by pushing the index switch push button. Hold the button down until the head stops moving.
32. Turn manual head position to "**Down.**"

CAUTION: Watch the movement of the core rods as the core rods move down, keeping a hand on the switch. If the core rods do not line up with the mold cavity correctly, turn the switch to the "**Up**" position **immediately**.

Then index the core rods to the stripper station and remove them. Change the .050 shims to smaller or larger ones. This is determined from what you saw by lowering the core rods near the neck ring pocket. The correct set-up for the core rod to cavity fit is .003" to .006" between core rod and the back seat of the mold neck ring. Refer to (Figure 9.2) and item 15 part C of the injection mold.

33. Re-install core rods and check as you did in step 32.

CAUTION: If the set-up is correct, go on to the next step. If not correct repeat step 32 until it is correct.

34. After shim size has been determined for the injection mold station, the same procedure should be carried out with the blow mold to match. Refer to step 16.
35. A) Rotate the index head until the side with the core rods in is at the stripper station.
     B) Remove the two core rods that you used for mold alignment. Leave core rod retaining plate off.
36. Rotate the index head to the next position and install the next face block (core rod holder). Then do the same for the remaining side of the index head. Make sure the trigger bars are in the full back position. Leave core rod retaining plates off. Remember to check that all cap screws are tightened.
37. Connect the water cooling hoses as indicated on the set-up sheet at the injection mold station or use this basic water connection drawing set-up. See Figure 9.3.
     A) All jumpers from the top half to the bottom half of the injection mold should be done nearest the operator. Hoses on the back side of the mold will be connected to manifolds, if supplied, or directly to the mold control unit if not manifold equipped.
     B) The inlet lines to the mold unit should always be connected to the bottom half of the mold. Return lines connected to the upper half of the mold. (Water flow pattern will be: inlet line in bottom back, jumpered to top front on operator's side, return line on top back ports.) This connection scheme purges any air trapped in the system.

CAUTION: Injection mold hoses are all stainless steel braid type only.

38. Alternate water/oil connection. Water/oil can be ported "**In**" in the top half and "**Out**" in the bottom half of the mold. Every other port of the same heat zone is alternating sides for "**In**" and "**Out**." The bottom half is connected the same way. This alternate way is sometimes used to equalize mold temperatures on both sides of the mold. It is especially helpful on very large molds.
39. Read the instructions on the mold control unit you are going to use. Put in air purge cycle. After purging air, turn the unit to cooling cycle and test for leaks. Correct leaks or if there are no leaks, set

temperatures at 150°F and let mold heat up. Check for leaks under heated conditions.

40. Connect cold water hoses to the blow mold. Make sure all hoses are tied out of the way of the indexing core rods.

41. Connect the cable to the thermocouple in the manifold on the parison die set.

42. Plug in manifold heaters.

43. Installation of core rods: Remove core rod retaining plate from face block at stripper station to install core rods. Check the trigger bar to insure it is in full back position. Insert core rod into face block. Check to see where retaining slot on core rod is in relation to the front surface of the face block. It should be recessed approximately .010" less than the face bar surface. To adjust core rods you need a special core rod tool supplied with machine. To adjust core rod "in" (move retainer slot toward head) loosen lock nut and turn star nut clockwise, then re-tighen locknut. Do this until the desired position, previously stated, is obtained. Repeat this procedure for all core rods.

    Install core rod retaining plates, tighten all retaining plate bolts. Place manual head switch to "down" position. When head reaches full down position, close hand valve for external tip cooling.

    Open hand valve for internal tip cooling and small hand valve for stripper station tip extension. Place selector switch for tip cooling in "manual" position. This will cause core rods to open and air to blow out of openings. Check core rod openings to insure they are all open approximately the same amount and that air is coming through. Turn tip cooling switch to "off." Adjust core rods as necessary, or if they are okay, proceed to next station and repeat this step for remainder of stations and core rods. A good starting point is about .070" gap.

44. Setting up stripper bar: Place manual head switch to "**Down**" position. When head reaches extreme down position, place stripper switch to "**Out**" position. This will cause stripper to move out. Install stripper bar in stripper assembly and **only hand tighten bolts**, so that bar can be slid left to right easily. Try, by looking, to center core rod in stripper bar contour. Place stripper switch to "off" position. Stripper will return to back position. Center contour holes on core rods and tighten stripper bar holding bolts. Check to see that stripper contour plate is approximately at place where bottle threads are to be molded on core rods. If it is not, loosen the four bolts that hold stripper assembly to guide shafts, and slide assembly in or out to reach the above position. Re-tighten holding bolts.

45. Connect external tip cooling hose to the stripper air connection.

46. Figures 9.3 and 9.4 show the component parts of the injection and blow mold.

Air assist mold die-set

CAUTION: The air assist mold die-set is a system built into the bottom die-set plate that allows you to move a large heavy mold and die-set easily. When turned on it will float the die-set and moid on a cushion of air.

CAUTION: The user of a die-set with Air Assist must be very careful and fully aware of how it reacts, as it not only floats the heavy mold and die-set but it can move very fast. **Extreme Caution** should be taken when using compressed air in the handling of die movement and the amount of air pressure used.

The air assist system incorporates a recessed cavity on the bottom half of the die-set. A source of compressed air is connected to the die-set which gives lift to the heavy weight, allowing it to float above the table.

CAUTION: It will move around very fast. Air pressure should be no more than **40 psl maximum**. The air line should have a regulator to evenly reduce this pressure and a **quick shut off** valve to control movement of the die-set. It moves fast and you must stop it fast. The compressed air line connection is at the rear of the die-set.

Mold cleaning

1. Turn on the chiller water to the blow mold and let it circulate through the mold. Turn on the injection mold temperature control unit letting it circulate through mold.
2. Remove the core rods at the stripper station. Check the valves for any foreign material that may be accumulated in the valve seat. Wash core rods in a solvent. Wash out core rod retainer block and top index head.
3. Turn on the hydraulic pump and index head to next station. Lower the head.
4. Block the platen open and shut off the hydraulic pump.
5. Clean the injection manifold and adaptor of all accumulated plastic. If more than the normal amount of plastic has built up between the manifold and mold, it is an indiction that:

    a. Nozzles are cracked.
    b. Nozzles are not seated in the mold properly.
    c. The nozzles are oval.
    d. Nozzle retainer bolts are broken or loose.
    e. Nozzle to manifold seat dirty or scored.

    In any case, excessive build-up of plastic in this area requires further investigation and corretive action be taken.

6. Remove all material which may be stuck between the mold halves. **Particular caution must be taken when using tools near the molds;**

use only brass or copper scrapers and use them carefully to prevent damage to the mold surfaces. Parting lines are extremely susceptable to damage.

7. Using a rag and solvent, wash out both the top and bottom mold halves of the parison mold. Pay close attention to the core rod seats and sealing surfaces. Corners in the core rod seats require special attention; be sure they are clean.

8. Check the mold for damage. Neck rings and part line are the more likely trouble spots. Make sure the neck rings are seated properly and are flush with the mold parting surface.

Look at some of the last containers made to see if any mold damage is evident.

NOTE: Neck rings should project .0005" above the mold part line.

9. Repeat step #2 with the core rods now at the stripper station.

10. Start the pump, raise and index the index head so the last group of core rods is at the stripper station. Lower the head and shut off the pump.

11. Again repeat step #2 with the core rods.

12. Now with rag and solvent wash out the blow mold, both upper and lower halves, paying particular attention to the vents and core rod seats.

13. Be sure no plastic is stuck between the mold halves. The bottom plug retainer screws may be a problem.

14. If movable bottom plugs are used, wash the slide mechanisms down. Check for any galling and if the plugs slide freely in and out. Re-lubricate slides and mating cam surfaces.

15. Lubricate both die-set guide posts.

16. Check the stripper plate, clean and wipe down.

17. Reinstall the core rods at the stripper station. Turn on the pump, raise the head, and index the head. Repeat the above for the next two stations. Also refer to the section of the manual on installing the molds for additional information on the core rod installation.

18. When all core rods are installed, remove the safety blocking under the top platens and lower the head, then close the molds so nothing falls into them. Shut the pump, power and water off.

Let Us Stress Again: Use Extreme Caution
This heavy mold will move very quickly and easily, but it also must be **stopped quickly**. It is also **very important** that the machine is perfectly level so the die-set and mold do not run off the table. The quick shut-off valve can be used to stop the mold in case arty problem should arise.

Balancing The Nozzle Manifold

This is an important process function and should be checked before starting a production run. Without the proper nozzle balance you can experience difficulty with wall distribution, consistency and dimensional stability.

A typical unbalanced manifold pattern would be experienced during a short shot as shown in Figure 9.6. Notice that the center mold cavity is filled more than either of the cavities on each side of it. During a full injection shot this condition might go unnoticed, but will eventually will create adverse molding conditions. A set-up man might try to correct these conditions with pressures and heats. This would only compound the problem with short shots, flash, folds and uneven wall distribution. It is evident that the plastic is taking the path of least resistance.

To correct this problem, the set-up man should remove the manifold and measure the nozzle orifice size. By increasing the diameter of the two outside cavities, you will increase the flow through them. Consequently, the cavities fill more uniformly. Figure 9.7 shows balanced injection mold cavities.

Figure 9.3 is a illustration of the manifold assembly and individual nozzles. The machine is now ready to start.

It is good practice to purge the plasticizer and check the actual melt temperature using a melt probe to know if the melt is at the temperature set on the temperature controllers. Usually, the actual melt of the extrudate will be higher than the actual temperature settings.

In the manual operation you should close the molds, shoot the material using approximately 80% of the desired shot size. You can then determine if all the cavities are filling the same. This is the same procedure followed in injection molding referred to as "short shooting" the mold. In the manual mode, strip the short shots and if the set-up looks correct, then set the proper shot size and turn the machine to auto for start up of production.

It is good practice to have the rear zone lower than the transition zone and the meter zone on a standard reciprocating screw plasticizer. The cushion should be approximate 1/4 inch (6.25mm) and either a ball check or a check ring should be used on the end of the screw.

The injection unit is usually set for ten (10) inches (245mm) of movement to fill an injection mold. Thus, when using velocity profiling there are ten (10) divisions for the shot size being used. If you are using three (3) inches (76.2mm) of the screw movement, then this three (3) inches (76.2mm) can be divided by ten (10) increments. The axiom of spooning material in the mold is a good one to follow. This refers to starting the flow in to the parison with low velocity and injection pressure to prevent core rod deflection. Once the melted homogenous plastic material is around the tip of the core rod and the core rod's end is now supported by melted material then ramp up to full injection pressure that you have chosen and fill the cavities, once filled then go into hold or pack pressure. This is usually 60% of the injection pressure being used. The hold or pack pressure keeps pressure on the parison, not allowing the parison to shrink. Once the injected heated plastic begins to solidify and its hydraulic pressure is equal to the hold or pack pressure you have set, you should then decompress the injection pressure and begin recovery of the screw.

When using a vertical plastifier, the above is not followed. In the vertical plastifier, the pressure is developed by the vertical screw forcing the melted homogenous material into the parison molds based on a predetermined screw run time and screw revolutions per minute. Once this time is exhausted, a small cylinder closes placing pressure on the melt, once again to prevent parison shrinkage.

In the vertical plastifier there is no way to use back pressure to assist in mixing of the plastic resin as in a reciprocating plastifier.

As a rule, vertical plastifiers are not used for injection blow molding of plastic resins that require high torque as PET, PC, ABS, polysulfone, nylon, etc.

The vertical plastifiers have wide usage in the manufacture of containers using the olefins and PVC.

chapter ten

The machine

Components

A normal three-station injection blow molding machine consists of a base, a horizontal platen, an injection clamp, a blow clamp, a stripper station, and a guarding plastic process control station. On a four-station machine, the extra station can used for safety checking the core rods or for conditioning the core rods with air, and the extra station could even have a small clamp station for this purpose (e.g., machines made by Alcan and Novapax). The stations on a four-station machine index 90 degrees, instead of the 120 degrees of machines made by JOMAR, Milacron, and Rainville. The only company that produces an all-electric machine is (ALCAN) Wheaton, and they design and have them built for their own use. Novapax uses an electric drive with hydraulic clamps. JOMAR, Milacron, and Rainville each use all-hydraulic systems.

Plastifiers

All the injection blow molding machines except the JOMAR vertical screw style machines use an injection unit with a reciproscrew. The reciproscrew offers the advantage in that it requires less purging, and color changes can be achieved with less loss of mixed resin. The injection pressure is applied directly to the melt and is not used to compress the unmelted pellets of resin, providing a fast rate of injection and good shot size uniformity.

Schematic D provides the nomenclature of a typical reciproscrew. The screw design should be adapted to the plastic resin that is being used.

The normal length to diameter (L/D) for the plasticating units is 28/1. However, injection blow molding machines may be produced with 20/1 lengths to diameter (L/D's) and may be as large as 30/1 length to diameter (L/D). It is up to the user to give the specifics for this feature to the machine manufacturer. This is also true for the screw compression, or else the supplier will provide what is referred to in the industry as a general purpose screw design. Compression ratios that are used for the olefins can be 3.5/1, 3.6/1,

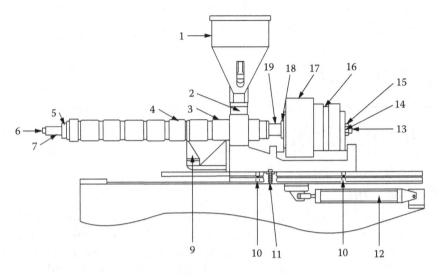

Figure 10.1 Schematic D Reciprocating screw nomenclature.

or 3.7/1. If you are running shear-sensitive resins, such as polyvinyl chloride or polyethylene terephthalate, then the screw compression could be possibly range from 2.2/1 up to 2.5/1. The screw compression ratio can be measured before the screw is installed in the plastifiers barrel. Through use of calipers or depth micrometers, measure the depth of the rear flight on the screw from its top down to the screws base diameter. Then measure the depth of the front flight located behind the screws tip. Measure from the top of the flight down to the screws base diameter. For example, if the rear screw flight measures one half inch and the front flight measures two tenths of an inch, then the compression ratio would be $0.5/0.2 = 2.5$ or the screws compression ratio would be 2 1/2.

You should always talk to your resin supplier and ask for their recommendations as to screw design, L/D of the plastifier, screw speed, and heating profile. Always keep in mind that the screw normally provides 90% of the heat to the melt. The heater bands have two functions. The first is to allow the user to start the machine from a cold start, and the second is to provide an even heat to the barrel so that the heat profile of the plastifier does not have peaks and valleys of cold or hot sections as you inject the melt into the parisons. Thus, the screw design is critical to quality production.

Index unit and interchangeable rotating table

The central unit of the injection blow molding machine is the precision backlash-free indexing unit, built by either CAMCO or FERGUSON. These units may rotate either 120° or 90° are designed to lift to a specific height and then index and drop down the same distance that it initially lifted. This height is usually 2–3 inches, depending on the machine.

The interchangeable rotating head is attached to the CAMCO or FERGUSON index drive by several bolts. The heads can be ordered in different thicknesses, although the normal thickness is 1.5 inches. It is best to use the 2-inch-thick rotating table so that an o-ring can be mounted between each face bar and the rotating table. Chapter 8, which describes the tooling, shows the face bar, rotating table, and so on, plus all tooling.

Horizontal platen

The horizontal platen is a steel plate that is machined and ground flat and that is usually 1.5–2.5 inches thick. The concern in injection blow molding is that the platen has to be thick enough and supported sufficiently so the parison injection mold and its die set, and the blow molds and its die set have minimum deflection. The horizontal platen has to be so designed to support the maximum number of parison molds that the machine can handle based on the length of the rotating tables trigger bar length. The platen should be oiled with a fine-grade machine oil and stoned to remover any scratches, rust, and so on. The horizontal platen will have a bolt pattern, to be used to bolt the injection parison tooling and the blow molding tooling. There is no standard bolt pattern for injection blow molding machines. Each manufacturer has its own bolt pattern layout.

There are two moving platens—one for the injection parison station and one for the blow mold station. Each has platen has its own specific bolt pattern for mounting the die set of the injection parison molds, and one for mounting the die set containing the blow mold.

Clamps

There are normally two clamp stations—one for the injection station and one for the blow mold station. The clamping forces are not equal. For instance, on a Milacron Rainville 88-ton injection blow molding machine, the injection clamp is rated at 75.5 tons and the blow clamp is rated at 12.5 tons, which equals 88 tons of total clamp forces. All injection blow molding machines use hydraulics for the clamps except the (ALCAN) Wheaton all-electric machine, which uses ball screws with servo drives.

Stripper station

The stripper station, or pick-up station, is mounted at the third index area. It is also hydraulic and has its own hydraulic pump and cylinder to push out. Through the use of air, it can tip 90 degrees to set the blown bottle onto a conveyor or into cartons. Some producers have a setup in which, as the containers are at the strip station, a flame treater will go over and back on the blown containers as they are held in the horizontal position for good flame treating; others flame-treat the blown containers as they pass on a conveyor downstream from the strip station. The strippers are also setup so that the producer can use air to cool the tips or body of the core rods once the blown containers are stripped from the core rods.

Guarding

Each manufacturer has its own design guards. Safety switches and locking devices engage when the guards are moved to the open position. These guards and safety interlocks should always remain in place and be checked at each setup for safety. All emergency stops should be checked daily to ensure that they are fully operational.

Process control

There are injection molding machines in use today that still have discrete controls or push buttons. However, most machines now have some type of process control. There are many different producers of process controls, and every producer has a preference. The process controllers in use are made by Barber Coleman, Allen Bradley, Hunkar, Siemens, General Electric, and other manufacturer brands or designs. Many process controllers control the screw speed, heater bands, injection pressure, hold pressure, back pressure, cycle time, injection delay, nozzle temperature, injection, blow delay, blow time (one or two stages), stripper delay, tip cooling, velocity profile of the injection unit, auto cycle, high and low injection pressure safeties, and high and low temperature settings for safety. These controllers can be programmed with a laptop computer, and the data can be locked in and retrieved. Usually, the process controller used has come from use on injection machines with added features for blow air and so on. Users usually choose a process controller with which their production personnel are familiar and that their maintenance personnel can troubleshoot. Usually, the process controller used on injection blow molding machines has been developed for use on injection molding machines. The injection blow mold machinery builders modify the process controllers logic so it can be used on their injection blow molding machines. They have to add features as blow delay, blow time; indexing of the rotating table, the stripper action, and exhaust time. Users of injection blow molding machines will usually specify what process controller they would want on their injection blow molding machine if this is an option from the injection blow molding machinery producer, to have a process controller with which three production personnel are familiar within use, they feel three maintenance personnel can troubleshoot and maintain, and that the area has several men for the process controller in their area such as suppled by Allen Bradley, Siemens, Hunkar, General Electric, etc.

Processing

Injection blow molding processing really begins with the parison design. Once you have a good parison design, then the tooling has to be designed to allow you to produce the parison and to blow the bottle. Key considerations before starting the machine or installing tooling are water, air, and electrics.

Waterlines from the tower are used to cool the feed throat of the injection unit and the heat exchange. Both waterlines should be made of 2-inch NPT and brought at least to 3 feet of the machine. The return lines should be of the same size. All the water should be treated for pH, alkalinity, iron, and so on. Compressed air should be oil-free, dry air. At least a 5-μ type filter should be installed before connecting to the blow molding machine.

Waterlines from a cooling tower are used to cool the feed throat and the heat exchanger on the injection blow molding machine.

Water supplied by either a central chillery system or from individual chillers is used to cool the blow molds. All the waterlines coming to the machine should be of the size of two (2) inch. National standard thread (NPT) and be brought as close to the machine waterlines as possible where the machinery producer or the mold builder has areas for connecting these water supply lines.

Airlines should be as large as possible but not under these quarters (3/4) inch in diameter with a national standard thread (NPT) coming to the machine, to connect to the areas provided by the injection blow molding machine builder. An air gauge should be installed if there is not one supplied by the injection blow molder machinery builder. This will ensure the operator that there is an adequate supply of blow on to blow the bottles or products. If the air gauge shows a drop of greater than five (5) pounds per square inch (psi), then the air supply is not adequate to produce or blow the product. If this happens, it is necessary to place a surge tank next to the machine so the blow air pressure does not drop greater than five (5) pounds per square inch (psi). This is particularly a problem in plants that have many injection blow molding machines producing blow molded bottles or where the air compressor is located away from the injection blow molding machines. Pressure drop occurs due to friction of the air traveling in piping to the machines, plus too small diameter piping is used to supply air to the blow molding machines. You should have at least one hundred anf fifty (150) pounds per square inch (psi) available at each blow molding machine. Many companies today are supplying their blow molding machines with a maximum of two hundred and fifty (250) pounds per square inch (psi).

The electrical connections are found in the injection blow molding manual for the particular machine and should only be connected by a licensed electrician.

It is wise to check the machine's operation in the manual mode before installing any tooling. Each check should include the following: stripper clamp for injection and blow, in open and close positions; table index; tip cooling; manual head cylinder; plastifier positioning cylinder; head on/off switch; plastifier screw rotation; injection; and relief.

On a new machine or a machine that has been rebuilt, the hydraulic pressure settings should be checked for each pump. The machine's manual will provide you with the number of pumps and their recommended settings. Once this check has been completed, close the safety gate and start the machine. Ensure proper functioning of all safety gate interlocks. (The operation must stop when opening the safety gate.) At the same time, verify the cushion of clamping cylinders so that they are not bottoming out within the cylinder and the clamps are smooth without any shock to the machine. The emergency stop (the large red button on the control panel) should stop all operations, including the electric pump motor.

You are now ready to install the tooling or do the mold setup. Using the recommended setup from your resin supplier, from a previous setup sheet, or from a setup that is stored in the process control memory, begin by turning the water to the feed throat and the heat exchanger. The hydraulics can now be turned off and the heat profile for the barrel, nozzle, and manifold be turned on and set for the desired temperature.

The rear zone or feed section of the screw should be set at a lower temperature than the work section or transition area of the screw. For instance, if HDPE is being processed, the rear heater bands could be set for 350°/360°F. The range of 350°F/360°F is given due to the fact there is no finite temperature to use, but this is a range from 350°F up to 360°F of temperature that is normally set by the operator or set-up personnel during set-up to get ready to start the machine and have the heats on the plastifier set to melt the plastic resin being processed. Normally the barrel heats will be at three set temperatures in 20 to 30 minutes. The plastifier should not be started until all the heat zones are the barrel of the plastifier are at their respective heat set points. Then the next heating zone could be set for 350°/360°F and the front zone, the nozzle, and the manifold for 390°/410°F. Normally, the heats will be up to temperature in approximately 20–30 minutes. Once this is achieved and the screw speed has been set for approximately 80–120 rpm with 200 psi back pressure, the injection pressure is set for 3000–3500 psi for high-pressure injection. It is a rule of thumb that the hold or pack pressure should be approximately 60% of the high-injection pressure. The hold or pack pressure is to prevent any unfill or sink in the parison. The hold pressure should be held until the hydraulic pressure of the plastic is equal to the hold pressure, and then the hold pressure should be dropped off and the screw decompressed. The screw will then back up and mix ready for the next shot size. The screw in the heated barrel will then back-up and as if backs up plastic pellets are conveyed along the screw from the feed zone of the screw, to the transition or work section of the screw, to the meter zone of the screw. All this time the plastic resin being conveyed turns from pellets to a water homogeneous melt. Thus the screw is ready to move forward and push out or shoot the next shot size to be used to produce the desired parison. Too much cushion will cause irregular shot size in the parisons. The cushion for the screw is usually set to be 0.25–0.375 inch. Never bottom the screw out against the end cap.

A possible setup for HDPE could be as shown below. Turn on the mold cooling setting the neck zone at 180°F, the body zone of the preform at 200°F, and the Possible Setup gate area at 180°F.

The timers could be set as follows:

1. Injection delay: 2.0 seconds
2. Injection: 5.0 seconds
3. Screw: 3.0 seconds
4. Blow delay: 2.0 seconds
5. Blow: 1.0 seconds
6. Stripper delay: 0.5 seconds
7. Tip cooling: 2.0 seconds

The shot size is strictly determined by the weight of the container and the number of mold cavities. The above numbers should be used as a starting point only if nothing else is available.

You should always purge the machine, using the purge shield. The extrudate should be checked for its actual temperature, using a melt probe. Each machine has its own characteristics, and the temperature settings may not represent the actual melt or extrudate temperatures; thus, reading the actual melt temperature will allow you to correlate actual melt temperature to what settings are used on the machine.

You should also become accustomed to seeing what a good homogeneous melt looks like coming out of the machine nozzle. There should be no spattering, jetting, or bubble, which can indicate moisture or screw wear. The melt should be clear if natural; if colorants are used, the colorant should not look degraded or have streaks or burn marks. This damage can be related to too-high heat settings, improper mixing, or too high a screw speed.

If you have a velocity profile available, it is wise to start to fill the parison cavity at a low velocity until the heated plastic has filled around the core rod tip, and then to switch to rapid injection to fill the cavity quickly. This method should minimize core rod deflection caused by high injection pressure. Core rod deflection will yield uneven parison walls, which in turn will cause uneven wall thickness, thin spots in the blown container, or possibly blowouts or tiny holes in the gate area or heels of the container.

One of the problems encountered by many injection blow molders is that the tooling purchased does not allow for a balanced runner, nor can each gate area be temperature controlled. All too often the temperature of the manifold is the only temperature control available. If the manifold used spherical seated nozzles, with each one having its own heater band, then if one secondary nozzle is drooling or stringing, its temperature could be dropped without affecting the other secondary nozzles or the manifold's melt. Log-style manifolds are the lowest-cost manifolds; however, in overall efficiency and lost time at the machine, they are the most expensive. The floor or production personnel should have the best tooling and have controls at their disposal to fine tune the process to achieve a total of 95%–98% quality product.

Once the quantity of product has been produced, the machine can be shut down. When the molds open, turn off the plastifier screw; move the injection unit away from the manifold; and allow all containers to be clear of the molds, stripper, and machine; then turn the machine to manual operation. Open the molds, making sure the safety blocks for the blow and injection clamps are functioning, and wipe them down while checking for any mold damage. Turn off the chiller to the blow molds. Lower the injection mold temperature unit having the injection mold open. Allow the temperature control units to stabilize to room temperature before closing the injection parison mold. The injection unit should be purged of all material with the feed hopper chute closed. Once this is complete, turn off the heater bands to the injection unit, then turn off the machine's main power and clean up the machine, noting any machine problems and keeping a complete set of the last containers plus a copy of the set up used for this production run.

chapter eleven

Maintenance

Injection molding and injection blow molding machines are very similar. The plasticating unit of both machines is very similar. Usually, both units use reciproscrew injection units and are run by hydraulics. The hydraulic oil should be changed any time the oil shows signs of a dark color (deep amber), which indicates oxidation. This is usually caused by overheating and burning of the oil. If the machine indicates hot oil or shuts off because of hot oil, the oil should be checked for oxidation. Change the oil immediately if it should become the color of coffee with cream, which indicates that water is in the oil. The heat exchanger should be checked for leakage or malfunction. Some plants use chilled water in the heat exchanger, which is fine if the chiller is above 42°F. Tower water is predominantly used; however, in areas in which daytime temperatures are over 95°F for an extended period of time, tower water may be too warm to remove heat through the heat exchanger, and the oil will overheat. The following images are of a maintenance checklist guide. If you have questions, consult the machine supplier and check the maintenance manual that was provided with your machine.

Maintenance Checklist

ITEM	FREQUENCY	ADD	CHANGE
GREASE ALL FITTINGS	SEE CHART		
HYDRAULIC OIL*	DAILY	YES	2000 HRS.
DRAIN AND CLEAN RESERVOIR	YEARLY		6000 HRS.
FILTER ELEMENTS	BI-MONTHLY		60 DAYS
EXTRUDER THRUST HOUSING (EGAN EXTRUDER)	WEEKLY	YES	2400 HRS.
INDEX UNIT	WEEKLY	YES	2400 HRS.
GEAR REDUCER	WEEKLY	YES	2400 HRS.
CLEAN HEAT EXCHANGER	6 MONTHS		
CHECK OUT ALL SAFETIES	WEEKLY		
CLEAN MACHINE	WEEKLY		
INSPECT FOR LEAKS	DAILY		
TIGHTNESS OF CLAMP TIE ROD NUTS	DAILY		
TIGHTNESS OF LIMIT SWITCH BOLTS	DAILY		
SAFETY GATE OPERATION	EACH SHIFT		
CLEAN MOLDS	WEEKLY		

* Note: Oil should be changed any time the oil shows signs of a dark color (deep amber), which indicates oxidation. Usually caused by over heating and burning of the oil.

Change oil immediately if the oil should become coffee with cream color. Water is in it. Check the heat exchanger.

Lubrication Location Chart

LOCATION OF FITTING	NUMBER OF FITTINGS	LUBE	FREQUENCY TO LUBE
BOTTOM BEARING BLOCK HUB BUSHING	1	EP-2	DAILY
CLAMP GUIDE RODS	4	EP-2	DAILY
POLYGON SHAFT (INDEX) SPLINE SHAFT	1	EP-2	DAILY
EXTRUDER THRUST BEARING HOUSING GUIDE ROD BUSHING (FLAT WEAR PAD FOR EGAN)	2	EP-2	DAILY
DIE-SET GUIDE POST	4	EP-2	DAILY
STRIPPER GUIDE RODS	2	EP-2	DAILY
INDEX UNIT GREAR BOX	4	BB	WEEKLY
HOT OIL/WATER DISTRIBUTOR (OPTIONAL)	1	EP-2	WEEKLY
EXTRUDER BOTTOM SLIDE	6	EP-2	MONTHLY
MAIN PUMP MOTOR BEARINGS	2	EP-2	6-MONTHS
RECIRCULATING PUMP MOTOR	2	EP-2	6-MONTHS
HYDRAULIC OIL		DTE-26	2000 HRS.
EGAN THRUST HOUSING	1	BB	2400 HRS.
INDEX UNIT HOUSING	1	46 SAE 90	2400 HRS.
GEAR REDUCER HOUSING	1	BB	2400 HRS.
PUMP COUPLINGS	2	EP-2	YEARLY

It is necessary that a good maintenance program be set up and followed. The most important parts are machine lubrication, hydraulic oil, gear oil and filter elements.

Hydraulic oil

Mobil DTE-26 A premium high grade oil with additives of anti-rust, anti-wear and anti-foamation.

Grease fittings

Mobilux EP-2 An unleaded lithium 12 EP grease for both anti-friction and plain bearings under wet or dry conditions.

Gear reducer

Mobil DTE-BB A high quality double Thrust bearing: inhibited oil designed for long oil life (Egan reciproscrew).

Index unit

Mobilube 46 (SAE 90) Gear lubricant.

Hydraulic oil

The initial supply of oil should be changed at the end of 2000 hours on the hour meter. When replenishing oil in the reservior always use the same brand of oil that is in the reservoir.

Never mix brands of oil

Each manufacturer makes certain claims, but the mix is not the same. Severe results are possible to occur when mixing two different brands together. Always flush, drain and clean the system of all oil first before using a different brand.

You can order an oil sampling kit where the oil may be tested for quality at any time. Please contact your machine supplier customer service department.

Weekly machine maintenance

1. Clean and remove all plastic pellets, scrap plastic and dirt that has accumulated around or on the machine. Clean under the machine also.
2. Wipe down the platens, index head and all flat surfaces, removing all oil, dust and dirt. Be sure to clean off the air probes under the index head.
3. Empty all drip pans; clean out all material which has fallen in them.
4. Clean the area around and under the machine both for safety and appearance.
5. Pick up all loose screws, nuts, or parts laying around and replace or put in a proper place.
6. Inspect the machine and repair any leak, guard, loose solenoid valve or connector.
7. Check the level of the hydraulic oil in the reservior. Add if required.
8. Test the operation of the safety gate system; check for any loose mounted limit switch or poor electrical connections.
9. Check lifting mechanism on index head.
10. Check trigger bar components, settings and operation.
11. Check the stripper assembly.
12. Check tie rod lock nuts for tightness.
13. Report any potential problem so your supervisor is aware and corrective action may be taken.

chapter twelve

Troubleshooting

In beginning any plastic processing, you should first analyze the problem. Once you have a good idea what you think is the problem, only one item should be changed at a time, and once that item has been changed, time should be allotted for the change to take effect. This allotted time may vary from a few minutes to a half hour or more. Once you have made this change to the setup, be sure to check that what you changed did not change something else in the process.

A list of various problems that occur in injection blow molding follows, with possible solutions. The list can be used as a general guide. It will not be all-inclusive, as new problems will occur with new materials.

Remember, you are injection molding parisons: If you make a good parison, you will produce a good product.

In the setup of the parison mold, a separate cooling line should be used for the neck rings, the end cap, and the cooling line above the end cap of the parison mold. In the blow mold, separate cooling lines should be used for the neck rings and the bottom plug. Normally, the product to be produced will dictate how many independent cooling lines should be used. It is good practice to have as many processing aids as possible so that you, the processor, can make individual changes without affecting the total parison or blow mold.

I recommend that all secondary nozzles have their own heater bands and thermocouples for good control. If they do not, then you have to raise the entire manifold temperature to get the one or two clogged nozzles to flow easier, yet the other nozzles do not need the higher temperature. If each of secondary nozzles do not have their own heater band, and one secondary nozzle is too hot and causing stringing of the heated plastic, the operator has to reduce the heat in the entire hot manifold to ease the problem. Yet, this problem was only with one secondary nozzle. If one of the secondary nozzles is freezing off, the operator may raise the temperature of the entire hot manifold. Yet, this problem was only affecting one of the secondary nozzles. In running the olefins as polypropylene (PP), high density polyethylene (HDPE), and low density polyethylene, raising or lowering the hot runner manifolds temperature may not be a problem. However, if you are running polyethylene terephthalate (PET), polyvinyl chloride (PVC), cyclic olefin copolymer (COC) or other fast setting resins that all shear sensitive or degrade due to high heat then raising the entire hot manifold temperature would not be acceptable.

The better the tooling is in design, the higher your efficiency will be in actual production, which is an everyday cost, whereas tooling expenses are a one-time charge. Thus, it is worth it to pay for good tool design and construction. The increased efficiency of your machine and quality of your product will more than offset the tooling cost.

Normally, machines are sold with a general purpose screw in the plastifier. You should always conduct a test to determine whether the plastifier can deliver a quality homogeneous melt at the output stated by the machine builder. If you are running engineering resins at PVC, PET, ABS, TPEs, polycarbonate, acrylics, or the new metalloncenes, it is best to purchase a screw for the specific resin you are using. Improper screw designs can cause nonhomogeneous melt, irregular shot sizes, bubbles, black specks, and short shots. There are many quality screw suppliers in the industry, and you should form an alliance with the one of your choice.

Finally, cycle time is money. You should know—for each machine—the actual time it takes to open, index, and close during the injection blow molding. Processing adds to the open, index, and close times. If you are competing with a machine that has a 3.2-second open, index, and close time and your competitor has machines that open, index, and close in 2.2 seconds, you are at a greater than 30% disadvantage. Larson Mardon Wheaton's new electric injection blow molding machine can open, index, and close in 1.1 seconds. For this and the reasons given previously, the four-station machine and the all-electric injection blow molding machine are best for cleanliness, quietness, and energy savings.

Table 12.1 Problems and Solutions

Short shots
    Out of material in the barrel
    Hopper out of material
    Material is bridging
    Material slide not open
    Material too cold
    Secondary gates dirty or orifice is not large enough
    Inadequate venting
    Shot size not adequate
Sink marks in parison
    Material not homogeneous
    Inadequate packing time or pressure
    Inadequate venting
Streaks in parison
    Mold dirty
    Regrind or material fines
    Cavity damaged
    Melt not homogeneous
    Injection pressure too high or too fast
Stringing of gate parison
    Melt temperature too high in secondary nozzle or manifold
    Secondary gate too large
Parison stuck to core rod
    Melt temperature too hot
    Parison mold coolant not adequate or at proper temperature
    Core rod cooling not adequate
Parison tip too large to compress
    Reduce land in secondary gate
    Move secondary nozzle in toward the parison
    Reduce land in end cap on spherical nozzles
Product torn
    Lower gate temperature
    Check core rods
    Check nozzle seats
    Check parting line of parison and blow mold
    Add injection time
    Replace secondary nozzle
Weak spot in center of product
    Lower gate temperature
    Check parison body temperature
    Lower parison body temperature
    Add more injection time and pack time
    Lower injection pressure
    Decrease back pressure
Heavy section in product
    Raise gate temperature
    Decrease core rod cooling
    Raise temperature in parison body

(Continued)

Table 12.1 Problems and Solutions

Push up not consistent
    Increase blow time
    Increase blow pressure
    Increase bottom plug cooling
    Reduce gate temperature
    Add core rod tip cooling
    Lengthen cycle time
    Check vents
Rocker bottoms
    Flash
    Vent
    Improper cooling
    Mold dirty
    Increase cycle time
    Check exhaust—possibly add exhaust time
    Check core rod openings
Bottom folds
    Increase blow pressure
    Check core rod openings
    Reduce injection pressure
    Increase temperature in parison mold at fold location
Surface finish
    Dirty molds
    Venting
    Material not homogeneous
    Temperature of parison too cold
    Increase blow pressure
    Increase cycle time
Dips in finish
    Parison not packed
    Vents
    Neck rings too cold
    Increase pack time
Cracked necks
    Raise melt temperature
    Increase parison neck ring temperature
    Retainer grooves on core rod too deep
Cocked necks
    Increase blow pressure
    Increase blow time
    Check bottom plug movement
    Increase cooling on blow mold body
Shrinkage
    Increase blow time
    Decrease blow mold temperature
    Increase pack pressure
    Increase pack time
    Lower parison mold temperature

(Continued)

Table 12.1 Problems and Solutions

Flash
Melt too hot
Injection pressure too high
Molds not flat
Vents too deep
Clamp not adequate
Platens not aligned
Mold damaged
Parison sag (parison parting line and blow mold parting line do not overlap)
Decrease melt temperature
Redo parison cooling lines for more balance
Venting
Add packing pressure
Melt not homogeneous
Check for nozzle uniform flow
Nozzle freeze off
Contamination
Damaged nozzle
Temperature too low
Manifold dirty
Thermocouple malfunctioning
Stripping
Decrease blow time
Retainer grooves too deep
Increase pressure to stripper
Lubricate stripper
Check core rods for damage
Weld lines
Raise manifold temperature
Raise parison mold temperature
Raise injection pressure
Increase pack time
Increase secondary nozzle orifice opening
Parting line
Molds damaged
Molds not aligned
Molds not flat
Clamp pressure inadequate
Check both blow mold and parison mold vents
Check blow mold and parison mold to determine which is the problem

chapter thirteen

Formulas

Determination of blow ratio

You should always know the blow ratio of your product and the normal or average thickness that the project requires. Localized areas can be lower than this nominal value if the parison contacts the wall sooner in different areas of the mold. Different materials or different grades of materials exhibit differing stretch behavior. The blow ratio of a molding is a way of representing the amount of stretch involved for a given combination of parison size and part size. For cylindrical containers, this can be expressed as

$$\text{Blow ratio} = \frac{\text{Mold diameter}}{\text{Parison diameter}}.$$

In general, this value is between 1.5 and 3, but it can be up to 5 in unusual cases.

Determination of average part thickness

The blow molded product should be designed in such a manner as to minimize any points of extreme stretching or too deep a draw, because of the associated stretch orientation and thinning. Surfaces should be as smooth as possible, and all corners should have generous radii. The ideal blow molded part design would be perfectly symmetrical, which would allow as even a wall thickness distribution as possible. However, other factors, such as aesthetics and function, play a part in design. The amount of stretching to which a parison is subjected is a function of the part size and configuration in relation to the parison size and orientation. In general, this can be expressed as follows:

$$\text{Average part thickness} = \frac{\text{Parison surface area}}{\text{Part surface area}} \times \text{parison thickness}.$$

Determination of the heat extraction load

The heat extraction load or the amount of heat to be removed from the product must be determined. This is important, as the amount of heat taken out by the blow mold must be known if the process is to be economically predictable. The amount of heat to be removed, Q, is determined by the material's temperature and the amount of plastic being delivered to the mold. It is calculated as follows:

$$Q = Cm\Delta t \ (0.003968),$$

where Q is the total change desired during molding, in British thermal units (BTUs; 0.003968 BTU = 1 calorie [constant]); C is the specific heat of the plastic material being processed, in calories per gram, in degrees Celsius; m is the amount of plastic per hour to be cooled expressed, in grams; and $\Delta t = (T_1-T_2)$, or the initial plastic (parison temperature into the mold) minus the final (demolding) temperature of the plastic, in degrees Celcius.

As an example, for determining the heat load for a typical mold, the following data are used:

- C, specific heat for polyethylene = 0.55 calories/gram, in degrees Celsius
- m, amount of PE to be cooled = 32 shots/hour × 18.75 pound/shot × 0.80 shot reduction length – 480 pounds/hour
- 1 pound = 453.5 grams [conversion factor]
- T parison temperature = 420°F (215.6°C)
- T demolding temperature = 100°F (37.8°C)
- Δt, material temperature change = $T-T$ = 215.6°C – 37.8°C = 177.8°C
- 1 calorie = 0.003968 BTU [constant]

Calculate the heat extraction load as follows:

$$Q = Cmt \ (0.003968);$$

$$= (0.55)(480 \times 453.59)(177.8)(0.003968);$$

$$= 84{,}483 \ \text{BTU/hour per mold, average.}$$

Assume 75% efficiency for heat transfer between chilled water and polyethylene. Then:

$$\text{Cooling required} = \frac{84{,}483}{0.75} = 112{,}644 \ \text{BTU/hr}$$

Reynolds number

With polyolefins, it is frequently desirable to run the molds as cold as possible, from 40° to 60°F (4.5°–15.5°C) or lower. Condensation or moisture on the mold can cause outside surface defects when the mold cooling temperatures are below the dew point. To reduce or eliminate these defects, it is possible to either increase the mold heat transfer fluid temperature or, using recently developed techniques, dehumidify the immediate blowing area to eliminate the condensation and maintain good surface appearance at a fast cycle. Effective dehumidification systems can be installed on existing equipment very satisfactorily, permitting ready access to the blow area while providing the dehumidification necessary to prevent condensation. Savings of 20%–30% have been reported through the use of this system. If the plant is not air conditioned, then on hot humid days, condensation will form on the blow molds. This moisture will be in the blow mold surface causing blemishes, ruptures, and thick sections all causing rejects. To prevent condensation from forming, the coolant supplied to the mold must have its temperature increased. When the mold coolant temperature is increased, the heated plastic entering the mold will not be cooled as rapidly, thus the overall cycle time will have to be increased. If the cycle time is increased, the number of parts produced per day drops and the product being produced uses in costs to be produced.

The use of existing water temperatures in plants can be supplemented to improve cooling conditions through the use of a turbulent flow of fluid through the mold channels. Depending on the channel sizes, the greater the volume and the higher the pressure of the fluid put through the channels, the greater the heat transfer. A turbulent flow wipes the side walls of the channels, permitting better heat transfer than is obtained with laminar flow of the fluid through the channels. This laminar flow characteristic tends to have reduced heat transfer at the channel circumference, whereas a turbulent flow characteristic tends to have reduced heat transfer at the channel circumference and enables more heat to be removed more quickly with higher fluid temperatures, which also reduces the possibility of condensation.

A large-capacity temperature control unit (e.g., with a 2–7.5-HP pumping capacity, depending on the mold and channel sizes) can not only provide the desire turbulent flow but also assist in maintaining uniformity of control throughout the mold on an automatic basis. Production reports indicate that temperature variations of no more than 1°–2°F (0.6°–1.1°C) are readily obtained with this approach, even with intricate molds.

To determine the proper flow for each mold, the flow's Reynolds number should be determined. The Reynolds number is a nondimensional parameter used to determine the nature of flow along surfaces. Numbers below 2100 represent laminar flow, numbers from 2100 to 5000 are transitional flow, and numbers above 5000 represent turbulent flow.

To determine the Reynolds number, N, the following calculations illustrate values derived for flows of 45 and 140 gallons per minute (gpm) and are based on a large-capacity temperature control unit controlling five zones per mold half of a blow mold:

$$N = (DV) \frac{P}{M},$$

where D is the pipe diameter (1 inch = 0.08333 foot), V is the fluid velocity in feet per second, P is the fluid density (72.3 pounds/cubic foot), M is the fluid viscosity (0.01002 poise × 0.0672 pound/foot-second per poise = 0.000673 pound/foot second).

$$V = \frac{\text{Flow rate, gpm}}{\text{No. of zones}} \times \text{Volume} \times \text{Time (in minutes)}/\text{Cross-sectional area of}$$
pipe;

$$\text{Volume of water} = 0.1337 \text{ cubit feet/gallon;}$$

$$\text{Time} = 1 \text{ second} = 0.1667 \text{ minute;}$$

$$\text{Cross-sectional area} = D0.5;$$

$$V \text{ for 45 gpm unit} = \frac{45}{5} (0.1337) \frac{1}{60} \frac{(0.08333)_2}{4} = 3.68 \text{ foot/second;}$$

$$V \text{ for 140 gpm unit} = \frac{140}{5} (0.1337) \frac{1}{60} \frac{(0.08333)_2}{4} = 11.44 \text{ foot/second;}$$

Then, the Reynolds numbers are:

$$N = DV = 0.08333 (3.68) = 32,944 \text{ for a 45-gpm unit,}$$

and

$$N = DV = 0.08333 (11.44) = 102,412 \text{ for a 140-gpm unit.}$$

Checklist

Injection blow molding product design

Designed By _____ Date _____

Customer _____ Quote No. _____

Date _____ Machine _____

Product Size _____ × _____ × _____ mm (inches)

Finish_____ mm   Snap Cap _____Other _____

Color _____ Let Down Cost _____

Color Manufacturer_____ Cost _____

Resin _____ Shrinkage _____

Gram Weight _____

Parting Line Location _____

Bottom Push-Up _____

    Product Finish   Gloss ☐ Matte ☐ Engraving ☐

    Annual Volume_____ Order Size _____

    Cavities _____ Estimated Cycle _____

                                    Actual Cycle _____

    Product Drawings _____ Date _____

    Approved by _____ Date _____

    Overflow Volume cc _____ ml _____

    Fill Point _____

    Average Wall Thickness _____

    Decorating Lug. Side _____ Bottom _____

    Silk Screen _____ Hot Stamp _____

    Offset _____ Label _____

    Pack Bulk _____ Layer _____

    Tool Vendor _____ Promised Date _____

    Preliminary Production Date _____ Approved Production Date _____

    Product Approved By _____

    Special Packing _____

    Product Specifications _____

    Product History _____

    Product Production Efficiency _____

    Engineering Comments _____

    Tooling Estimated Cost _____

    Tooling Actual Cost _____

    Tool Maintenance Cost _____

    Production Comments _____

Injection blow molding tooling design

Product _____ Product Drawing _____

End User _____

Tool Shop_____

P.O. Number _____ Date _____

Plastic Resin _____ Shrinkage _____

Date Quote _____ Number of Cavities _____

Date Promised_____

Machine Used _____

Colorant Producer _____ Let Down _____

Regrind   Yes ☐   No ☐

Machine Data _____

Make _____ Screw Type _____

Injection Clamp Tonnage _____ Blow Clamp Tonnage _____

Trigger Bar Length _____ inches _____

Dry Cycle _____ seconds _____

Rotating Table Thickness _____ inches _____

O-ring on Rotating Table   Yes ☐   No☐ Three Places ☐ Four Places ☐

Conditioning Station Yes ☐ No ☐

Process Control Yes ☐ No ☐

Type Control   Maco Siemens ☐   Allen Bradley ☐   Other ☐

Maximum Opening 10 inches _____ 12 inches _____ Special _____

Rotary Union Yes ☐ No ☐

Die Sets   Size _____

Make Own ☐ DME ☐ Progressive ☐ Mold Base Inc. ☐ Other _____

Preliminary Production Date _____ Approved Production Date _____

Product Approved By _____

Special Packing _____

Product Specifications _____

Product History _____

_____

_____

Product Production Efficiency _____

Engineering Comments _____

Tooling Estimated Cost _____

Tooling Actual Cost _____

Tool Maintenance Cost _____

Production Comments _____

_____

_____

_____

chapter fourteen

Typical injection blow molding machines and general information

In the United States there are two companies that produce injection blow molding machines. They are Milacron Uniloy and JOMAR.

Milacron purchased the Rainville machine in the 1990s and it is now marketed as Milacron Uniloy out of Manchester, Michigan. They offer various sizes of machines ranging from thirty-five ton up to one hundred eighty ton. They all are three station machines and have a horizontal recriposcrew as the injection unit. Both the clamps and the injection unit use hydraulics. They recently built an injection blow molding machine in Italy that offers an electric drive for the injection unit and hydraulics for the clamp.

JOMAR in southern New Jersey offers a line of injection molding machines ranging from fifteen ton up to approximately a two hundred ton. You can purchase the machines with either a vertical injection unit or a horizontal injection unit depending on the size of the machine. The injection unit and the clamps all use hydraulics.

Bekum in Berlin, Germany offers a seventy five ton, three station machine. It has a horizontal recriposcrew injection unit and the injection unit and the clamps all use hydraulics.

Novapax in Germany offers a four station sixty-five ton machine. It uses hydraulics for the clamp and electric drive for the injection unit.

Tri-Delta in New Jersey offers a three station machine. They are very similar to the Milacron Uniloy machines in size. They use hydraulics for the clamp and the injection unit.

Meccanoplastica of Italy offers the only all electric machine in the industry. It is called the JET 60. The number sixty indicates the clamp size. Both the clamps and the injection unit use electric for the clamp movement and the screw movement. This should save any company at least fifteen (15%) percent of their electric power consumption. Injection mold machine builders are all offering total electric machines due to using energy costs and the injection blow molding machines plus extrusion blow molding machine builders will all follow this energy saving trend.

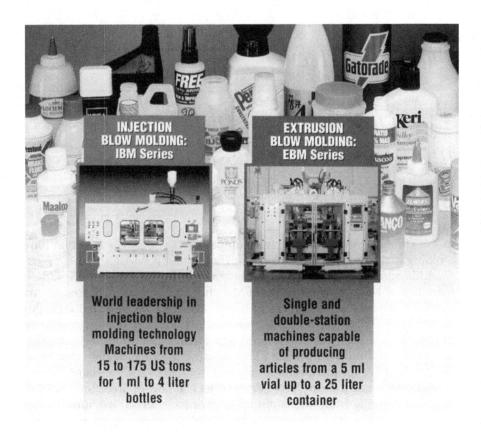

Figure 14.1 Courtesy of JOMAR —Vertical Screw 3 Station.

Thermal Controlled Core Rod Set-up

Figure 14.2 Courtesy of Uniloy Milacron 3 Station.

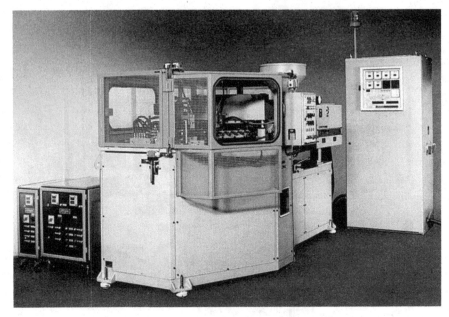

Figure 14.3 Courtesy of Bekum 3 Station.

Figure 14.4 Courtesy of Tri-Delta 3 Station.

Figure 14.5 Courtesy of Novapax. Station machine, Germany.

Reed Prentice

Figure 14.6 Reed Prentice.

Figure 14.7 Reed Prentice.

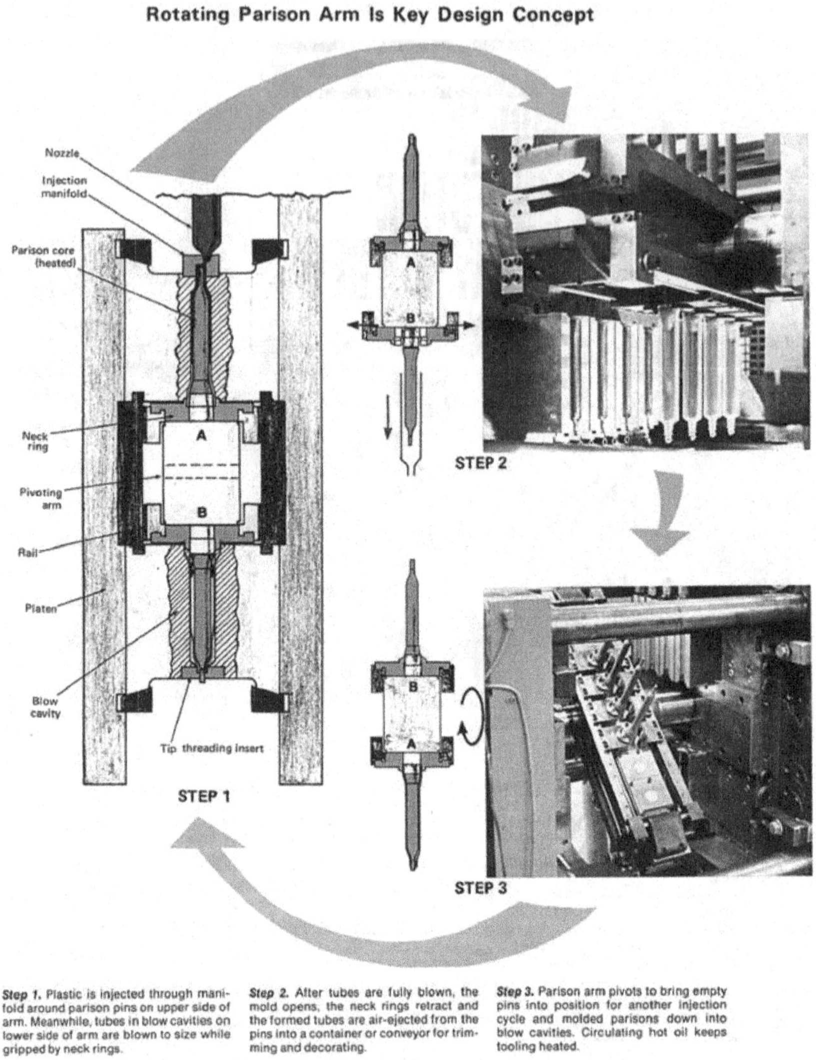
Rotating Parison Arm Is Key Design Concept
Nozzle
Injection manifold
Parison core (heated)
Neck ring
Pivoting arm
Rail
Platen
Blow cavity
Tip threading insert
A
B
STEP 1
STEP 2
STEP 3
Step 1. Plastic is injected through manifold around parison pins on upper side of arm. Meanwhile, tubes in blow cavities on lower side of arm are blown to size while gripped by neck rings.
Step 2. After tubes are fully blown, the mold opens, the neck rings retract and the formed tubes are air-ejected from the pins into a container or conveyor for trimming and decorating.
Step 3. Parison arm pivots to bring empty pins into position for another injection cycle and molded parisons down into blow cavities. Circulating hot oil keeps tooling heated.

Figure 14.8 Reed Prentice.

Figure 14.9 Reed Prentice.

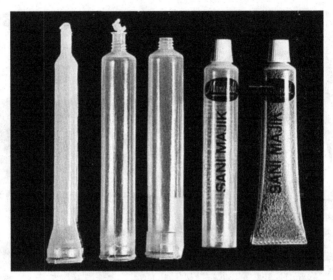

Figure 14.10 Reed Prentice.

Figure 14.11 Reed Prentice.

Figure 14.12 Reed Prentice.

Figure 14.13 All-electric injection blow molding machine. Meccanoplastica SRL, Italy.

Bibliography

"Injection Blow Molding Is Ready to Go," Modern Plastics (1970)

"Injection Blow Molding," Chester Strohecker, USI; Plastics Design & Processing (1980)

"Injection Blow Molding Rigid PVC Containers," Glen R. Ver Hage, Rainville Hoover Universal and Henry Nyman, Kerr Glass Corporation (1984)

"Fundamentals of Injection Blow Molding," Ralph Abramo, R. J. Abramo Associates, Society of Plastics Engineers Meeting (1986)

"Injection Blow Molding Bottles," J.R. Deeps, Captive Plastics; *Plastics Engineering* (34) (1975)

"Check Out All Options Before Picking a Process," J. Nancekwell, *Canadian Plastics* (44) (1986)

"Vertical Layout in JOMAR Range," JOMAR Industries, *Plastic Rubber Weekly* 1167 (1986)

"Uniloy Puts Its Technology on Show," Johnson Controls; *Plastic Rubber Weekly* 1167 (1986)

"Plastic Processing Machines," G. Krantz, Kunstoffe (1986)

"How Big is Blow Molding," Ann Brockschmidt, *Plastic Technology* 12 (1986), pp. 77–9

"Check Out All Options Before Picking A Process," J. Nancekwell, *Canadian Plastics* 44 (1986), pp. 36–40

"Injection Blow Molding—A Review," Emery I. Valyi, PhD. - Packaging

"New Developments in Injection Blow Molding," William B. Niemi, Reed Prentice (1970)

Injection blow molding

Injection clamp force tons kN	Blow clamp force tons kN	Shot capacity oz. g	Number of blowing stations	Platen size (length x width) in. mm	Swing radius in. mm	Max. day-light in. mm	In-mold trim (Y/N)	Special features	Floor space ft² m²	Weight lb. kg (000)	Model	Supplier[a]
12	3	1.2 PVC; 1.76 PE	1	11.25 x 10	14	12		3-station; vertical plastifier; programmable controls			15V	Jomar
22.4	2.24	3				4	N	3-station, 8 cavities; 350 ml. max.; PE, PP, PS, PC, PVC, and others			SBM-320	Bekum
30	2	1.2	2	12 x 6		2.75	Y	Thin wall tubes; low tooling costs	21	2.4	DUO 30	Ossberger-FGH
38	6.6	1.2, 2.6 PVC; 3, 6 PE	1	15.5 x 17.5	22.5	12		3-station; vertical plastifier; programmable controls; 2-pressure inject; 2-pressure blow			40V	Jomar
39.6	9.9	12 340	1	18 x 14.9 457 x 378	15.9 404	15 381		Microprocessor with full process control, 2-pressured blow, tie-bar clamp design	81 7.52	12.3 5.6	45-3S	Uniloy Milacron
44	3.3	2	2	15 x 8		4	Y	Thin wall tubes; low tooling costs	27	2.7	DUO 40	Ossberger-FGH
44.8	5.6	10	1	22 x 12		5	N	3-station, 12 cavities; 1000 ml. max.; horiz. press			SBM-341	Bekum
52	17	2.6, 3.6 PVC	1	17 x 24.5	25.5	14		3-station; vertical plastifier; programmable controls			65V	Jomar
61.6	15.4	14.8 420	1	23.6 x 14.9 600 x 378	18.15 461	15 381		Microprocessor with full process control	98 9.1	14.6 6.6	70-3S	Uniloy Milacron
72	17	5.2, 7.9 PVC; 6, 9.5, 12.6 PE	1	17 x 24.5	25.5	15		3-station; vertical plastifier; programmable controls; 2-pressure inject; 2-pressure blow			85-S V	Jomar
72.8	17.6	14.8 420	1	29.5 x 18	22.5 571	15 381		Microprocessor with full process control	137.4	22 10	85-3S	Uniloy Milacron
75.4	12.5	5.3 150	1	28 x 16	19.75 501	15 381		Microprocessor with full process control	101.4	23.5	88-3[b]	Uniloy Milacron
85	15 135	36 PS 950	1	18 x 36	32	12	Y	4 stations, stretch blow	82	18	85	B&G Products
88	4	4	1	16 x 8		6	Y	CVS boots, rack and pinion boots			SBE 50/140	Ossberger-FGH
88	12.5	8.8 250	1	30 x 22	28 61	15 381		Microprocessor with full process control	133 12.3	35 15.9	108-4	Uniloy Milacron
104	18.8	8.8 250	1	38 x 22	27.75 705	15, 21		Microprocessor with full process control	138.9	25.5	122-3[b]	Uniloy Milacron
111	28	35, 60 gms/ sec.	1	19 x 35.5	34.5	16		Capable of liquid control cores; programmable controls			115R	Jomar
112	19.3	21	1	39 x 24	32	6.3	N	4-station, 12 cavities; 1500 ml. max.; PE, PP, PS, PC, PET, PVC			SBM-4100	Bekum
114	18.8	13.8 390	1	45 x 22	31.5 800	15, 21		Microprocessor with full process control	142.5	27.5	135-3S[b]	Uniloy Milacron
115.7	27 245	18.3 520	1	48 x 21	31.4 798	16 406		Microprocessor with full process control	186.6	33.1 15	129-3S	Uniloy Milacron
132	8	5	2	16 x 8		6	Y	Steering wheel boots with side vent	62	6.79	SB 2/60	Ossberger-FGH
135	28	12.6, 15.8 PE 35, 60 gms/ sec.	1	24 x 35 19 x 35.5 19 x 35.5	34.5	16, 21		3-station; vertical plastifier; programmable controls; Programmable controls; dual-stage blow; 36-in. transfer head			135V 135R	Jomar
160	28	12.6, 15.8, 19.4 PE 35, 60 gms/ sec.	1	24 x 35 23 x 37.5	35 34.5	21		3-cylinder parison clamp; Programmable controls; dual-stage blow; 36-in. transfer head			160V 160R	Jomar
175	32	15.8, 19.4 PE	1	24 x 35 19 x 35.5	35	21		3-cylinder parison clamp			175V	Jomar
180	18.8	13.8 390	1	45 x 22	31.5 800	15, 21		Microprocessor with full process control	164 15.2	28 12.7	189-3S[b]	Uniloy Milacron
185	45 405	48 PS 1,300	1	24 x 48 600 x 1,200	36 900	12 300 18 450	Y	4 stations, stretch blow	180 17	29 13.1	185	B&G Products
	1.7	2-34	2	180 X 100		60	Y	Low mold costs, highly flexible machine	7	1.084	DUO 30	Ossberger
	3	6-50	2	200 X 150		90	Y	Low mold costs, highly flexible machine	8	1.22	DUO 40	Ossberger
	1.7	neto 140	1	210 x 160		140	Y	Low mold costs, highly flexible machine	17	1.58	SBE 50/140	Ossberger
	1.7	neto 140	1	210 x 160		140	Y	Low mold costs, highly flexible machine	20	1.65	DSE 50	Ossberger
	4	neto 140	2	210 x 160		140	Y	Highly flexible machine	24	2.5	SB2/60	Ossberger
	1	2-34	1	180 x 100		60	Y	Highly flexible machine	7	0.89	EHS 30	Ossberger
	1.5	6-50	1	200 x 150		90	Y	Highly flexible machine	7	0.97	EHS 40	Ossberger

a—For full names and addresses of suppliers, see the Buyers' Guide
b—Only available in North America

Index

A

air lines, 127
air pressure, 6
(ALCAN) Wheaton, 2, 14–15, 20, 123, 125
average part thickness, formula for
    determining, 141

B

Bekum machines, 2, 147
Beryellium copper (BeCu), 47
blenders, 17
blow molds, 79–81
    mounting, 80
    overview, 80
    setup, 79
    tooling, 80–81
blowability, 53
blow ratio, formula for determining, 141
blow-up ratio (BUR), 25, 31, 45
bottom pinch off, 21
bottom plug, 21, 88–94
    cost and, 89
    depth, 89
    disadvantages, 89
    drop impact and, 90
    explained, 88
    olefins and, 88
    rigid materials and, 88
bottom push-up. *See* bottom plug
BUR, 25, 31, 45

C

CAD, 29–31
CAMCO, 124–125
cavity numbers, 8

cellulose, 11
childproof closures, 7
clamps, 5–6, 8, 125
clarity, 21
COC, 66
coextrusion, 23
colorants, 15–16
    carrier resins and, 15
    crystallization and, 16
    melt index and, 15
    PET and, 15–16
    titanium dioxide, 16
compression ratios, 14
computer-aided design (CAD), 29–31
conditioning, 6
Constant Velocity Joint (CVJ), 39, 50
containers
    basic design, 29
    blow-up ratio, 25
    computer-aided design (CAD) and,
      29–31
    displacement, 29
    L/D ratio, 25
    oval, 24, 25, 39
    overflow capacity, 26, 29, 30
    parison shaping and, 39
    perform molds and, 25
    requirements, 20
    shapes, 25, 26–28
    wall thickness and, 31–32
core rods, 1–3
    air passages and, 46
    blow-up ratio and, 45
    body, 47
    bottom-opening and, 43
    components, 46–47
    designing, 43–45
    diameter, 32, 45–46
    flash chrome, 48

RELATED TITLES

Absorbable and Biodegradable Polymers
by Shalaby W. Shalaby
ISBN# 0849314844

Introduction to Polymer Science and Chemistry: A Problem Solving Approach
by Manas Chanda
ISBN# 0849373840

Polymers for Dental and Orthopedic Applications
by Shalaby W. Shalaby
ISBN# 0849315301

Plastics Technology Handbook, Fourth Edition
by Manas Chanda
ISBN# 0849370396

Practical Injection Molding
by Bernie Olmsted
ISBN# 0824705297

Computer-Aided Injection Mold Design and Manufacture
by Jerry Fuh
ISBN# 0824753143

Printed in the United States
by Baker & Taylor Publisher Services